Algebra, Logic and Combinatorics

LTCC Advanced Mathematics Series

Series Editors: Shaun Bullett *(Queen Mary University of London, UK)*
Tom Fearn *(University College London, UK)*
Frank Smith *(University College London, UK)*

Published

Vol. 2   Fluid and Solid Mechanics
edited by Shaun Bullett, Tom Fearn & Frank Smith

Vol. 3   Algebra, Logic and Combinatorics
edited by Shaun Bullett, Tom Fearn & Frank Smith

Forthcoming

Vol. 1   Advanced Techniques in Applied Mathematics
edited by Shaun Bullett, Tom Fearn & Frank Smith

LTCC Advanced Mathematics Series - Volume 3

Algebra, Logic and Combinatorics

Editors

Shaun Bullett
Queen Mary University of London, UK

Tom Fearn
University College London, UK

Frank Smith
University College London, UK

 **World Scientific**

NEW JERSEY • LONDON • SINGAPORE • BEIJING • SHANGHAI • HONG KONG • TAIPEI • CHENNAI • TOKYO

Published by

World Scientific Publishing Europe Ltd.
57 Shelton Street, Covent Garden, London WC2H 9HE
Head office: 5 Toh Tuck Link, Singapore 596224
USA office: 27 Warren Street, Suite 401-402, Hackensack, NJ 07601

Library of Congress Cataloging-in-Publication Data
Names: Bullett, Shaun, 1967– | Fearn, T., 1949– | Smith, F. T. (Frank T.), 1948–
Title: Algebra, logic, and combinatorics / [edited by]
    Shaun Bullett (Queen Mary University of London, UK),
    Tom Fearn (University College London, UK),
    Frank Smith (University College London, UK).
Description: New Jersey : World Scientific, 2016. |
    Series: LTCC advanced mathematics series ; volume 3
Identifiers: LCCN 2015049552| ISBN 9781786340290 (hc : alk. paper) |
    ISBN 9781786340306 (pbk : alk. paper)
Subjects: LCSH: Algebra. | Logic, Symbolic and mathematical. | Combinatorial analysis. |
    Differential equations.
Classification: LCC QA155 .A525 2016 | DDC 510--dc23
LC record available at http://lccn.loc.gov/2015049552

British Library Cataloguing-in-Publication Data
A catalogue record for this book is available from the British Library.

Desk Editors: R. Raghavarshini/Mary Simpson

Typeset by Stallion Press
Email: enquiries@stallionpress.com

Printed in Singapore

Preface

The *London Taught Course Centre (LTCC) for PhD students in the Mathematical Sciences* has the objective of introducing research students to a broad range of advanced topics. For some students, these topics might include one or two in areas directly related to their PhD projects, but the relevance of most will be much less clear or even apparently non-existent. However, all of us involved in mathematical research have experienced that extraordinary moment when the penny drops and some tiny gem of information from outside one's immediate research field turns out to be the key to unravelling a seemingly insoluble problem, or to opening up a new vista of mathematical structure. By offering our students advanced introductions to a range of different areas of mathematics, we hope to open their eyes to new possibilities that they might not otherwise encounter.

Each volume in this series consists of chapters on a group of related themes, based on modules taught at the LTCC by their authors. These modules were already short (five two-hour lectures) and in most cases the lecture notes here are even shorter, covering perhaps three-quarters of the content of the original LTCC course. This brevity was quite deliberate on the part of the editors: we asked contributors to keep their chapters short in order to allow as many topics as possible to be included in each volume, whilst keeping the volumes digestible. The chapters are "advanced introductions", and readers who wish to learn more are encouraged to continue elsewhere. There has been no attempt to make the coverage of topics comprehensive. That would be impossible in any case — any book or series of books which included all that a PhD student in mathematics might need to know would be so large as to be totally unreadable. Instead, what we present in this series is a cross-section of some of the topics, both classical and new, that have appeared in LTCC modules in the nine years since it was founded.

The present volume covers the general area of algebra, logic and combinatorics. The main readers are likely to be graduate students and more experienced researchers in the mathematical sciences, looking for introductions to areas with which they are unfamiliar. The mathematics presented is intended to be accessible to first year PhD students, whatever their specialised areas of research, though we appreciate that how "elementary" or "advanced" any particular chapter appears to be will differ widely from reader to reader. Whatever your mathematical background, we encourage you to dive in, and we hope that you will enjoy reading these concise introductory accounts written by experts at the forefront of current research.

Shaun Bullett, Tom Fearn, Frank Smith

Contents

Chapter 1

Enumerative Combinatorics

Peter J. Cameron

School of Mathematical Sciences,
*Queen Mary University of London, London E1 4NS, UK**
pjc20@st-andrews.ac.uk

This chapter presents a very brief introduction to enumerative combinatorics. After a section on formal power series, it discusses examples of counting subsets, partitions and permutations; techniques for solving recurrence relations; the inclusion–exclusion principle; the Möbius function of a poset; q-binomial coefficients; and orbit-counting. A section on the theory of species (introduced by André Joyal) follows. The chapter concludes with a number of exercises, some of which are worked.

1. Introduction

Combinatorics is the science of arrangements. We want to arrange objects according to certain rules, for example, digits in a sudoku grid. We can break the basic question into three parts:

- Is an arrangement according to the rules possible?
- If so, how many different arrangements are there?
- What properties (for example, symmetry) do the arrangements possess?

Enumerative combinatorics provides techniques for answering the second of these questions.

Unlike the case of sudoku, we are usually faced by an infinite sequence of problems indexed by a natural number n. So if a_n is the number of solutions to the problem with index n, then the solution of the problem is a sequence $(a_0, a_1, \ldots)$ of natural numbers. We combine these into a single

---

*Current address: School of Mathematics and Statistics, University of St Andrews, North Haugh, St Andrews KY16 9SS, UK.

object, a formal power series, sometimes called the *generating function* of the sequence. In the next section, we will briefly sketch the theory of formal power series.

For example, consider the problem:

Problem 1. *How many subsets of a set of size n are there?*

Of course, the answer is 2^n. The generating function is

$$\sum_{n\geq 0} 2^n x^n = \frac{1}{1-2x}.$$

Needless to say, in most cases we cannot expect such a complete answer!

In the remainder of the chapter, we examine some special cases, treating some of the important principles of combinatorics (such as counting up to symmetry and inclusion–exclusion).

An important part of the subject involves finding good asymptotic estimates for the solution; this is especially necessary if there is no simple formula for it. Space does not permit a detailed account of this; see Flajolet and Sedgewick [4] or Odlyzko [10].

The chapter concludes with some suggestions for further reading.

To conclude this section, recall the definition of the binomial coefficients:

$$\binom{n}{k} = \frac{n(n-1)\dots(n-k+1)}{k(k-1)\dots 1}.$$

A familiar problem of elementary combinatorics asks for the number of ways in which k objects can be chosen from a set of n, under various combinations of sampling rules:

	Without replacement	With replacement
Order significant	$n(n-1)\dots(n-k+1)$	n^k
Order not significant	$\binom{n}{k}$	$\binom{n+k-1}{k}$

2.   Formal Power Series

2.1.   *Definition*

It is sometimes said that formal power series were the 19th-century analogue of random-access memory.

Suppose that $(a_0, a_1, a_2, \dots)$ is an infinite sequence of numbers. We can wrap up the whole sequence into a single object, the *formal power series*

$A(x)$ in an indeterminate x given by

$$A(x) = \sum_{n \geq 0} a_n x^n = a_0 + a_1 x + a_2 x^2 + \cdots .$$

We have not lost any information, since the numbers a_n can be recovered from the power series:

$$a_n = \frac{1}{n!} \frac{\mathrm{d}^n}{\mathrm{d}x^n} A(x) \bigg|_{x=0}.$$

Of course, we will have to think carefully about what is going on here, especially if the power series doesn't converge, so that we cannot apply the techniques of analysis.

In fact, it is very important that our treatment should not depend on using analytic techniques. We define formal power series and operations on them abstractly, but at the end it is legitimate to think that formulae like the above are valid, and questions of convergence do not enter. So operations on formal power series are not allowed to involve infinite sums, for example; but finite sums are legitimate. The "coefficients" will usually be taken from some number system, but may indeed come from any commutative ring with identity.

Here is a brief survey of how it is done.

A *formal power series* is defined as simply a sequence $(a_n)_{n \geq 0}$; but keep in mind the representation of it as a formal sum $\sum a_n x^n$. Now:

- Addition and scalar multiplication are defined term-by-term:

$$\left(\sum a_n x^n \right) + \left(\sum b_n x^n \right) = \sum (a_n + b_n) x^n,$$

$$c \left(\sum a_n c^n \right) = \sum (c a_n) x^n.$$

- Multiplication of series is by the *convolution rule* (mysterious in the abstract, but clear in the series representation)

$$\left(\sum a_n x^n \right) \left(\sum b_n x^n \right) = \sum c_n x^n,$$

where

$$c_n = \sum_{k=0}^{n} a_k b_{n-k}.$$

- Differentiation of series (which will be denoted by D rather than $\mathrm{d}/\mathrm{d}x$) is term-by-term, using the rule that $\mathrm{D}(x^n) = n x^{n-1}$:

$$\mathrm{D} \left(\sum_{n \geq 0} a_n x^n \right) = \sum_{n \geq 1} n a_n x^{n-1} = \sum_{n \geq 0} (n+1) a_{n+1} x^n.$$

Note that, in the rule for product, the expression for c_n is a finite sum.

With the above addition and multiplication, the set $R[[x]]$ of all formal power series over a commutative ring R with identity is a commutative ring with identity. The third operation makes it a *differential ring*. This just says that differentiation is R-linear and that *Leibniz' law*

$$D(AB) = A(DB) + (DA)B,$$

holds.

Other operations on formal power series are possible. For example, we can form infinite sums and products, provided these only involve finite sums and products of coefficients. For example, if $(a_n^{(i)})$ are sequences with the property that, for any n, there exists m such that $a_n^{(i)} = 0$ for all $i > m$, then we can form $\sum_{i \geq 0} A_i$, where $A_i = \sum_{n \geq 0} a_n^{(i)} x^n$. For the coefficient of x^n in this infinite sum is the *finite* sum

$$\sum_{i=0}^{m} a_n^{(i)}.$$

We can substitute a formal power series $B(x)$ for x in another formal power series $A(x)$ provided that the constant term of B is zero. For the series $B(x)^i$ has the coefficients of $1, x, x^2, \ldots, x^{i-1}$ all zero; so by the preceding paragraph,

$$A(B(x)) = \sum_{i \geq 0} a_i B(x)^i$$

is well-defined.

You are invited to formulate a sufficient condition for the infinite product $\prod A_i(x)$ to be defined.

A formal power series $A(x)$ in $R[[x]]$ is invertible if and only if its constant term a_0 is invertible in R. To see this, consider the equation

$$\left(\sum a_n x^n \right) \left(\sum b_n x^n \right) = 1.$$

The constant term shows that $a_0 b_0 = 1$, so it is necessary that a_0 is invertible. But if a_0 is invertible, then the equation for the coefficient of x^n is

$$\sum_{k=0}^{n} a_k b_{n-k} = 0,$$

so that

$$b_n = -a_0^{-1} \left(\sum_{k=1}^{n} a_k b_{n-k} \right),$$

so b_n can be found recursively as a linear combination of $b_0, \ldots, b_{n-1}$.

An important special case is: a formal power series with constant term 1 is invertible (and its inverse also has constant term 1).

2.2. *Classical examples*

If the coefficients of a formal power series are numbers (as they will almost always be), then the series may or may not converge for a particular value of x. Recall from complex analysis that, for any power series with complex coefficients, there is a number $R \in [0, \infty]$ (a non-negative real number or infinity) with the properties

- If $|x| < R$, then $\sum a_n x^n$ converges;
- If $|x| > R$, then $\sum a_n x^n$ diverges.

We say that R is the *radius of convergence*; the behaviour of the series for $|x| = R$ is not specified. The interpretation of the extreme values is that, if $R = \infty$, then the series converges for all x, while if $x = 0$, then the series diverges for all $x \neq 0$.

If a series has non-zero radius of convergence, then it defines a complex analytic function inside its circle of convergence. This gives us several more techniques that can be used. For example,

- We can use Cauchy's integral formulae to evaluate the derivatives at the origin (the coefficients of the series);
- If some identity between power series is known for analytic reasons, then it holds in the ring of formal power series.

There are three very important series:

- The *exponential series*

$$\exp(x) = \sum_{n \geq 0} \frac{x^n}{n!}.$$

(We usually write $\exp(x)$ rather than e^x.)
- The *logarithmic series*

$$\log(1 + x) = \sum_{n \geq 1} \frac{(-1)^{n-1} x^n}{n}.$$

- The *binomial series*, for any complex number a:

$$(1 + x)^a = \sum_{n \geq 0} \binom{a}{n} x^n,$$

where $\binom{a}{n}$ is the *binomial coefficient*

$$\binom{a}{n} = \frac{a(a-1)\dots(a-n+1)}{n!}.$$

Note that the binomial series has only finitely many terms if a is a non-negative integer, since in that case if $n > a$ then the numerator of the binomial coefficient $\binom{a}{n}$ contains a factor $a - a = 0$. However, in all other cases, it is an infinite series.

Now various familiar properties hold, for example:

- $\exp(\log(1+x)) = 1+x$, $\log(1 + (\exp(x) - 1)) = x$ (the series $\exp(x) - 1$ has constant term zero and so can be substituted into the logarithmic series);
- More generally, $\exp(a\log(1+x)) = (1+x)^a$;
- The laws of exponents hold, for example $(1+x)^a(1+x)^b = (1+x)^{a+b}$, or $((1+x)^a)^b = (1+x)^{ab}$. (For the second, we have to write $(1+x)^a = 1 + A(x)$ for some power series $A(x)$.)

As said above, all these facts have analytic proofs, and therefore hold for the power series. However, we have the possibility of either finding proofs of the identities by combinatorial manipulations, or alternatively, of unpacking the combinatorial content of the equations to prove combinatorial identities.

Here is a simple example. Consider the identity

$$(1+x)^a(1+x)^b = (1+x)^{a+b}.$$

Calculating the coefficient of x^n on both sides, and using the formula for multiplication of formal power series, we obtain our first example of a *binomial coefficient identity*:

Theorem 1 (Vandermonde convolution).

$$\sum_{k=0}^{n} \binom{a}{k}\binom{b}{n-k} = \binom{a+b}{n}.$$

An even simpler example is the one discussed in the introduction:

$$(1-2x)^{-1} = \sum_{n\geq 0}(2x)^n,$$

an example of a geometric series.

2.3.   *Generalisations*

The simple notion of formal power series described here can be extended
in several ways. Here are a few:

- We will often represent a sequence (a_n) by its *exponential generating
  function*

$$\sum_{n \geq 0} \frac{a_n x^n}{n!}.$$

  We will see that this arises naturally in counting "labelled" objects.
- A *Laurent series* can admit a finite number of negative powers of x:

$$A(x) = \sum_{n \geq n_0} a_n x^n,$$

  where n_0 may be negative. Of course, we could allow arbitrary negative
  as well as positive powers of x; but then the sum in the convolution rule
  for the product of two series would be infinite, so this doesn't work.
- We can consider formal power series in more than one variable, expressions of the form

$$A(x, y) = \sum_{m, n \geq 0} a_{m,n} x^m y^n.$$

  We see that a formal power series in two variables is the generating function for a two-dimensional array of numbers. Nothing new is required,
  since $A(x, y)$ belongs to the ring of formal power series in the indeterminate y over the ring $R[[x]]$.
- We can even allow infinitely many indeterminates (as long as each term
  only involves finitely many of them).
- There are other completely different kinds of series. Number theorists like
  the *Dirichlet series* which represents a sequence $(a_n)_{n \geq 1}$ by the series

$$\sum_{n \geq 1} \frac{a_n}{n^s}.$$

There is a framework which includes both kinds of series, but that is
beyond the scope of this chapter.

3.   Subsets, Partitions, Permutations

We have met very little combinatorics yet. The most important combinatorial objects are subsets, partitions, and permutations, and these provide many counting problems, to which we now turn.

3.1.   *Subsets*

The Binomial Theorem for non-negative integers n states:

Theorem 2 (Binomial Theorem).

$$(1+x)^n = \sum_{k=0}^{n} \binom{n}{k} x^k.$$

This is a polynomial in x. Substituting $x = 1$, we obtain

$$2^n = \sum_{k=0}^{n} \binom{n}{k}.$$

The left-hand side of this equation is the number of subsets of the set $\{1, 2, \ldots, n\}$. On the right-hand side, we use a familiar interpretation of the binomial coefficients: $\binom{n}{k}$ is the number of k-element subsets of $\{1, 2, \ldots, n\}$. Since every subset has a unique cardinality in the range $0, \ldots, n$, we see why the equation is true. Indeed, once we have verified the counting interpretation of the binomial coefficients, we have given a bijective proof of the Binomial Theorem for non-negative integer exponents. (The term "bijective proof" refers to an argument which shows that two expressions are equal by finding a bijection or matching between sets counted by the two expressions.)

There is a huge industry of finding and verifying *binomial coefficient identities*. In the preceding section, we met the Vandermonde convolution (written here with different variables)

$$\sum_{l=0}^{k} \binom{n}{l} \binom{m}{k-l} = \binom{n+m}{k}.$$

Here is a bijective proof. Take a class of $n + m$ children, of whom n are girls and m are boys. We wish to pick a team made up of k of the children in the class. This can obviously be done in $\binom{n+m}{k}$ ways. Alternatively, we could choose a number l between 0 and k, and select l of the n girls, and $k - l$ of the m boys. For any given l, there are $\binom{n}{l}\binom{m}{k-l}$ ways to do this. Summing over l gives the result since each selection occurs once in this process.

Let us find a two-variable generating function for the binomial coefficients, regarded as a two-parameter family. Sums will be over all relevant values of the indices. Since

$$\sum \binom{n}{k} x^k = (1+x)^n,$$

we have

$$\sum\sum \binom{n}{k} x^k y^n = \sum_n ((1+x)y)^n = (1 - y - xy)^{-1},$$

the required generating function.

Writing the right-hand side as

$$(1-y)^{-1}\left(\frac{1-xy}{(1-y)}\right)^{-1} = (1-y)^{-1}\sum_k \left(\frac{xy}{1-y}\right)^k$$

and taking the coefficient of x^k on both sides, we see that the generating function for binomial coefficients with k fixed and n varying is

$$\sum_{n\geq 0} \binom{n}{k} y^n = \frac{y^k}{(1-y)^{k+1}}. \tag{1}$$

This can be rearranged to form a proof of the Binomial Theorem for negative exponents. Note that, for $n = 0$, we have the geometric series

$$\sum y^n = \frac{1}{1-y}.$$

3.2. *Partitions*

A *partition* of a set X is a collection of non-empty subsets of X which are pairwise disjoint and have union X. The *Bell number* $B(n)$ is defined to be the number of partitions of $\{1, 2, \ldots, n\}$.

There is no simple formula for $B(n)$, but its values can be found recursively. Suppose that the part containing the element n has size k. Then it consists of n together with a $(k-1)$-element subset of $\{1, \ldots, n-1\}$ $\left(\text{which can be chosen in } \binom{n-1}{k-1} \text{ ways}\right)$, and the remaining $n - k$ points can be partitioned in $B(n-k)$ ways. So

$$B(n) = \sum_{k=1}^{n} \binom{n-1}{k-1} B(n-k).$$

Multiplying both sides by $x^{n-1}/(n-1)!$ and summing over n, we see that the *exponential generating function* $\beta(x) = \sum B(n)x^n/n!$ satisfies

$$\frac{d}{dx}\beta(x) = e^x \beta(x),$$

from which (with $\beta(0) = 1$) we find $\beta(x) = \exp(\exp(x) - 1)$.

Despite this simple form for the generating function, there is no simple formula for the Bell numbers themselves, and even finding a good asymptotic estimate is surprisingly hard.

Just as the binomial coefficients refine the total number 2^n of subsets of an n-set, we can define $S(n,k)$ to be the number of partitions of $\{1,\ldots,n\}$ having k parts. So $B(n) = \sum_k S(n,k)$. The numbers $S(n,k)$ are the *Stirling numbers of the second kind.* The analogue of the Binomial Theorem is the equation

Theorem 3.

$$\sum_{k=1}^{n} S(n,k)(x)_k = x^n.$$

Here, $(x)_k$ denotes the "falling factorial" $x(x-1)\ldots(x-k+1)$ $\left(\text{the numerator of } \binom{x}{k}\right)$.

3.3.   *Permutations*

The number of permutations of the set $\{1,\ldots,n\}$ is given by the factorial function $n!$ Its ordinary generating function does not converge (however, see Exercise 5); but its exponential generating function is $(1-x)^{-1}$.

Again we can refine this, as follows. Any permutation of $\{1,\ldots,n\}$ can be written uniquely as the composition of cyclic permutations on disjoint subsets. (We must count cycles of length 1 here.) The *unsigned Stirling number of the first kind*, written $u(n,k)$, is the number of permutations which are products of exactly k disjoint cycles (including fixed points). It is well known from algebra that we can associate a *sign* with a permutation; if π has k cycles, then the sign of π is $\text{sgn}(\pi) = (-1)^{n-k}$. Then the *signed*

Stirling number of the first kind is $s(n, k) = (-1)^{n-k}u(n, k)$; in other words, we count each permutation with the appropriate sign.

The Stirling numbers of the first kind satisfy

Theorem 4.

$$\sum_{k=1}^{n} s(n, k)x^k = (x)_n.$$

Comparing this with the equation for the Stirling numbers of the second kind, we see that these numbers are the coefficients of the change of basis for the space of polynomials between the basis of powers of x and the basis of falling factorials and back again. Hence, the infinite lower triangular matrices whose entries are the Stirling numbers of the (signed) first and second kind are inverses of one another.

4. Recurrence Relations

A *recurrence relation* is simply a definition of the nth term of a sequence in terms of n and the preceding terms in the sequence. A simple example is the sequence of *factorial numbers*, defined by

$$0! = 1,$$
$$n! = n \cdot (n - 1)! \quad \text{for } n > 0.$$

A simple induction proof shows that there is a unique sequence defined by this prescription.

We saw already that there is a very close connection between recurrence relations and inverting formal power series. This material is hopefully familiar from elementary combinatorics, so the treatment will be brief. I will take a single example of each type of recurrence I discuss.

4.1. *Linear, finite length, constant coefficients*

The standard famous example here is the *Fibonacci sequence*, partially defined by the famous recurrence

$$F_n = F_{n-1} + F_{n-2} \quad \text{for } n \geq 2.$$

It is clear that once values for F_0 and F_1 are given, the entire sequence is determined. Be warned that there are several conventions for what the "official" initial values are!

Let $\Phi(x) = \sum F_n x^n$. The recurrence relation shows that the coefficient of x^n in $(1 - x - x^2)\Phi(x)$ is zero for $n \geq 2$; so

$$\Phi(x) = \frac{c + dx}{1 - x - x^2}$$

for some constants c, d. These constants can be found from the values of F_0 and F_1: explicitly, $c = F_0$ and $d = F_0 + F_1$. Then the technique of partial fractions can be used to find an explicit formula for F_n. Write

$$\Phi(x) = \frac{a}{1 - \alpha x} + \frac{b}{1 - \beta x} = \sum_{n \geq 0} (a\alpha^n + b\beta^n) x^n,$$

where $(1 - \alpha x)(1 - \beta x) = 1 - x - x^2$; so $F_n = a\alpha^n + b\beta^n$.

There are two natural places where Fibonacci numbers arise:

- In how many ways can n be written as a sum of ones and twos (with order important)? If a_n denotes this number, then $a_0 = a_1 = 1$ and $a_n = a_{n-1} + a_{n-2}$ for $n \geq 2$.
- How many sequences of 0s and 1s contain no two consecutive 1s? If this number is b_n, then we have $b_0 = 1$, $b_1 = 2$ and $b_n = b_{n-1} + b_{n-2}$ for $n \geq 2$.

In general, any linear recurrence of finite length with constant coefficients, say

$$a_n = \sum_{i=1}^{k} c_i a_{n-i} \quad \text{for } n \geq k$$

has generating function of the form $p(x)/(1 - \sum_{i=1}^{k} a_i x^i)$, where p is a polynomial with degree smaller than k. (The polynomial $1 - \sum_{i=1}^{k} a_i x^i$ is called the *characteristic polynomial* of the recurrence.) Using partial fractions gives an explicit formula. If

$$1 - \sum_{i=1}^{k} a_i x^i = \prod_{i=1}^{r} (1 - \alpha_i x)_i^d,$$

where the α_i are distinct, then the solution is a linear combination of the sequences $(n^j \alpha_i^n)$ for $i = 1, \ldots, r$ and $j = 0, \ldots, d_i - 1$; the k initial values determine the coefficients.

4.2. *Linear, finite length, polynomial coefficients*

There is a general technique for solving recurrence relations of the form

$$a_n = \sum_{i=1}^{k} p_i(n) a_{n-i},$$

where p_i are polynomials in n, rather than constants. Rather than an expression for the generating function as a rational function, we obtain a differential equation for it with rational functions as coefficients. Here is a simple example.

Problem 2. *In how many ways can we partition $\{1, 2, \ldots, n\}$ into subsets of sizes 1 and 2?*

Let s_n denote this number. Then $s_0 = s_1 = 1$, and

$$s_n = s_{n-1} + (n-1)s_{n-2}$$

for $n \geq 2$ (the two terms on the right corresponding to the cases that n lies in a singleton or doubleton part). Let $S(x)$ be the exponential generating function $\sum s_n x^n / n!$. Multiplying the recurrence by $x^{n-1}/(n-1)!$ and summing, we see that

$$\frac{\mathrm{d}}{\mathrm{d}x} S(x) = (1+x) S(x).$$

With initial condition $S(0) = 1$, this has solution $S(x) = \exp(x + x^2/2)$.

4.3. *Other linear recurrence relations*

Another very important recurrence relation arose in the work of Euler on partitions of integers. A *partition* of n is an expression for n as a sum of positive integers in non-increasing order. Let $p(n)$ denote the number of partitions of n, and $P(x) = \sum p(n)x^n$ its ordinary generating function. It is straightforward, by equating powers of x, to show that

$$P(x) = \prod_{k \geq 1} (1 - x^k)^{-1}.$$

So the coefficients of the inverse function $\prod(1 - x^k)$ will be the coefficients in a recurrence relation for the partition numbers.

Now an argument similar to the previous one shows that

$$\prod_{k \geq 1} (1 - x^k) = \sum_{n \geq 0} q(n)x^n,$$

where $q(n)$ is the number of partitions of n with an even number of parts all distinct, minus the number of partitions with an odd number of parts, all distinct. This number is given by *Euler's Pentagonal Numbers Theorem*:

Theorem 5 (Euler's Pentagonal Numbers Theorem).

$$q(n) = \begin{cases} (-1)^k & \text{if } n = k(3k-1)/2 \text{ for some integer } k, \\ 0 & \text{otherwise.} \end{cases}$$

So we obtain the recurrence

$$p(n) = \sum_k (-1)^{k-1} p\left(\frac{n - k(3k-1)}{2}\right),$$

where the sum is over all non-zero integers k (positive and negative) for which the argument of p on the right is non-negative. Thus,

$$p(n) = p(n-1) + p(n-2) - p(n-5) - p(n-7) + p(n-12) + \cdots.$$

This recurrence allows efficient computation of the partition numbers; indeed, a table computed by McMahon was used in the investigations of Hardy, Littlewood and Ramanujan on partition numbers.

4.4. *Nonlinear recurrence relations*

It is impossible to survey this category in general, so here is one important example, the *Catalan numbers*.

Problem 3. *In how many ways can we bracket a non-associative product of n factors?*

Let B_n be the number of bracketings. So, for example, $B_3 = 2$, the two bracketings being $(x \circ (y \circ z))$ and $((x \circ y) \circ z)$. We have $B_1 = B_2 = 1$, and we take $B_0 = 0$ by convention. Let $f(x)$ be the ordinary generating function.

A bracketing of n factors is obtained by splitting the factors into the first k and the last $n - k$ (for some k with $0 < k < n$), bracketing these, and combining. So we have a quadratic recurrence relation

$$B_n = \sum_{k=1}^{n-1} B_k B_{n-k}$$

for $n \geq 2$. This equation says that $f(x)^2$ agrees with $f(x)$ on all powers of x except the first; so

$$f(x) = x + f(x)^2.$$

This equation for $f(x)$ is quadratic, and can be solved to give

$$f(x) = \tfrac{1}{2}\left(1 \pm \sqrt{1 - 4x}\right).$$

The fact that $f(0) = 0$ forces us to take the minus sign.

Then we can expand using the Binomial Theorem with exponent $\tfrac{1}{2}$. Finally we obtain a remarkably simple formula for the nth *Catalan number* $C_n = B_{n+1}$:

Theorem 6.

$$C_n = \frac{1}{n+1}\binom{2n}{n}.$$

Note that it is not at all obvious *a priori* that $n + 1$ divides the binomial coefficient $\binom{2n}{n}$.

The Catalan numbers are one of the most ubiquitous sequences in combinatorics. Stanley's book [13] contains many different occurrences of these numbers. I will mention just one here. The number of rooted binary trees with n leaves is the Catalan number C_{n-1}. The specification of these trees is

- There is a binary tree consisting only of a root;
- Any other binary tree T consists of an ordered pair (L, R) of binary trees (with their roots joined to the root of T).

It is clear from this specification that, if T_n is the number of rooted binary trees with n leaves, then $T_1 = 1$ and

$$T_n = \sum_{k=1}^{n-1} T_k T_{n-k}$$

for $n \geq 2$; so $T_n = B_n = C_{n-1}$ for all n.

Figure 1 shows the five trees with four leaves.

Fig. 1.   Binary trees with four leaves

5.   Inclusion–exclusion

5.1.   *The principle*

Let $A_1, A_2, \ldots, A_n$ be subsets of a set X. For any subset I of the index set $\{1, 2, \ldots, n\}$, let

$$A(I) = \bigcap_{i \in I} A_i,$$

be the set of elements lying in all sets with index in I (and possibly others), and $B(I)$ be the set of elements lying in all sets with index in I but not in any sets with index not in I. By convention, $A(\emptyset) = X$.

It is easy to calculate the cardinalities of the sets $A(I)$ in terms of the sets $B(I)$:

$$|A(I)| = \sum_{J \supseteq I} |B(J)|.$$

The *inclusion–exclusion principle* allows us to go the other way:

Theorem 7 (Inclusion–exclusion principle).

$$|B(I)| = \sum_{J \supseteq I} (-1)^{|I|-|J|} |A(J)|.$$

To see this, count the contribution of each element $x \in A(I)$ to the right-hand side. Suppose that x belongs to k sets A_j for $j \notin I$. If $k = 0$, then x counts once in $A(I)$ and not in any other sets. If $k > 0$, then it is easy to see (using the Binomial Theorem) that the positive and negative contributions of x cancel. So on the right we are counting just the points in $B(I)$, as required.

The most important special case is where $I = \emptyset$: the number of points lying in none of the sets A_i is

$$\sum_{J} (-1)^{|J|} |A(J)|,$$

where the sum is over all subsets J of $\{1, \ldots, n\}$.

5.2.   *Two examples*

Two counting questions are easily solved using this principle.

Problem 4. *How many surjective functions are there from an n-set to a k-set?*

Let X be the set of all functions from $\{1,\ldots,n\}$ to $\{1,\ldots,k\}$, and let A_i consist of those for which i is not in the image. Then surjective functions are those lying in none of the sets A_i. We have $|A(I)| = (k-|I|)^n$, and so the required number of functions is

$$\sum_{J\subseteq\{1,\ldots,k\}} (-1)^{|J|}(k-|J|)^n = \sum_{j=0}^{k}(-1)^j \binom{k}{j}(k-j)^n.$$

Note that this number is $k!\,S(n,k)$, where $S(n,k)$ is the Stirling number of the second kind.

Problem 5. *How many derangements (fixed-point-free permutations) of an n-set are there?*

Let X be the set of permutations of $\{1,\ldots,n\}$, and A_i the set of permutations which fix the point i. Then derangements are permutations lying in none of the sets A_i. We have $|A(I)| = (n-|I|)!$, and so the required number of derangements is

$$\sum_{J\subseteq\{1,\ldots,n\}} (-1)^{|J|}(n-|J|)! = \sum_{j=0}^{n}(-1)^j \binom{n}{j}(n-j)!.$$

Using this, it is easy to show that the exponential generating function for the number of derangements is $(1-x)\exp(-x)$.

5.3.   *Chromatic polynomial*

The next example is of importance both in graph theory and in statistical mechanics: it was invented by Birkhoff in an attempt to prove the Four-Colour Conjecture (as it was then).

A *graph* consists of a set V of *vertices* and a set E of 2-element subsets of V called *edges*. (Note that our definition forbids loops and multiple edges.) A graph is *connected* if, for any pair of vertices, there is a sequence of vertices starting at the first and ending at the second, each consecutive pair forming an edge. In general, a graph is uniquely the disjoint union of *connected components*.

A *proper colouring* of a graph with q colours $1,\ldots,q$ is a function from the vertex set to the set of colours so that two vertices forming an edge form different colours.

Birkhoff showed that the number of colourings of a graph Γ with q colours is the evaluation at q of a monic polynomial whose degree is the number of vertices, called the *chromatic polynomial* of Γ.

To see this, let X be the set of all colourings (proper or improper) of the vertices of Γ with q colours. For each edge $e \in E$, let A_e be the set of colourings for which the ends of e have the same colour. Then the proper colourings are precisely those lying in none of the sets A_e.

For a subset I of E, we need to count the colourings for which the edges in I (and possibly others) are "bad" (the ends have the same colour). Consider the graph (V, I) with edge set I. If all edges in I are bad, then any two vertices which lie in the same connected component have the same colour; so $A(I)$ is the number of assignments of a colour to each connected component of (V, I), which is $q^{c(I)}$, where $c(I)$ denotes the number of connected components of (V, I).

By inclusion–exclusion, the number of proper colourings is

$$\sum_{I \subseteq E} (-1)^{|I|} q^{c(I)}.$$

This is clearly a polynomial. The term of largest degree occurs for the graph with most connected components, which is uniquely the case $I = \emptyset$ and gives the term q^n (where $n = |V|$).

6. Posets and the Möbius Function

The Möbius function of a partially ordered set is a generalisation both of the inclusion–exclusion principle and the classical number-theoretic Möbius function, and has many applications.

6.1. *Definitions*

A *partially ordered set*, or *poset* P is a pair $(A, \leq)$, where A is a set and $\leq$ a binary relation on A which is

- *Reflexive*, that is, $a \leq a$ for all a;
- *Antisymmetric*, that is, $a \leq b$ and $b \leq a$ imply $a = b$;
- *Transitive*, that is, $a \leq b$ and $b \leq c$ imply $a \leq c$.

An *interval* in a poset is a set of the form $[a, b] = \{x : a \leq x \leq b\}$. A poset is *locally finite* if every interval is finite. The theory below applies to any locally finite poset.

Examples of posets include:

- The *Boolean lattice* $B(n)$, consisting of all subsets of $\{1, \ldots, n\}$, ordered by inclusion.
- The *partition lattice*, consisting of all partitions of $\{1, \ldots, n\}$, ordered by refinement.
- The set of positive integers, ordered by divisibility.
- The *direct product* of posets is defined on the Cartesian product of the underlying sets, with the order defined coordinatewise.

Let $P = (A, \le)$ be a locally finite poset. The *incidence algebra* of P over a field F (usually taken to be the rational numbers) is the set of functions $f : A \times A \to F$ with the property that $f(a, b) = 0$ unless $a \le b$. Addition and scalar multiplication are defined pointwise, and multiplication is given by

$$(fg)(a, b) = \sum_{x \in A} f(a, x) g(x, b).$$

Note that the only non-zero terms in the sum arise when $x \in [a, b]$, and so the sum is finite and well-defined.

Three elements of the algebra of particular importance are:

- The *identity* ι, given by

$$\iota(a, b) = \begin{cases} 1 & \text{if } a = b, \\ 0 & \text{otherwise.} \end{cases}$$

- The *zeta function* ζ, given by

$$\zeta(a, b) = \begin{cases} 1 & \text{if } a \le b, \\ 0 & \text{otherwise.} \end{cases}$$

- The *Möbius function* μ, the inverse of ζ (with respect to the above multiplication). It satisfies

$$\mu(a, a) = 1, \qquad \sum_{x \in [a, b]} \mu(a, x) = 0 \quad \text{if } a < b.$$

The last equation allows μ to be calculated recursively:

$$\mu(a, b) = - \sum_{a \le x < b} \mu(a, x) \quad \text{if } a < b.$$

For example:

- In the Boolean lattice $B(n)$, if $a \le b$, then $\mu(a, b) = (-1)^{|b \setminus a|}$.

- In a chain (a totally ordered set),

$$\mu(a,b) = \begin{cases} 1 & \text{if } a = b, \\ -1 & \text{if } b \text{ covers } a \text{ (i.e., } [a,b] = \{a,b\}). \\ 0 & \text{otherwise.} \end{cases}$$

- In the positive integers ordered by divisibility,

$$\mu(a,b) = \begin{cases} (-1)^r & \text{if } b/a \text{ is the product of } r \text{ distinct primes,} \\ 0 & \text{otherwise.} \end{cases}$$

The important property of the Möbius function, generalising the inclusion–exclusion principle (Theorem 7), is the following.

Theorem 8 (Möbius inversion theorem). *Let P be a locally finite poset, and let f, g belong to the incidence algebra of P. Then the following two statements are equivalent:*

- $g(a,b) = \sum_{a \leq x \leq b} f(a,x);$
- $f(a,b) = \sum_{a \leq x \leq b} g(a,x)\mu(x,b).$

In many interesting cases, the poset P has the "homogeneity property" that there is a least element 0 and, for any $a < b$, then there exists c such that the intervals $[a, b]$ and $[0, c]$ are isomorphic. It follows that $\mu(a, b) = \mu(0, c)$, and so the Möbius function of such a poset can be regarded as a one-variable function.

6.2. *Gaussian coefficients*

The *Gaussian* coefficients generalise binomial coefficients by introducing an extra parameter which is traditionally called q: these numbers are sometimes called *q-binomial coefficients*.

They are defined as follows, for non-negative integers n and k:

$$\begin{bmatrix} n \\ k \end{bmatrix}_q = \frac{(q^n - 1)(q^{n-1} - 1)\ldots(q^{n-k+1} - 1)}{(q^k - 1)(q^{k-1} - 1)\ldots(q - 1)}.$$

Here are a few of their properties.

- $\lim_{q \to 1} \begin{bmatrix} n \\ k \end{bmatrix}_q = \begin{pmatrix} n \\ k \end{pmatrix}.$

- Consider lattice paths in the plane from the origin to the point (m, n), where each step in the path is one unit to the right or upwards. It is easy to see that the number of such paths is $\binom{n}{k}$. If we weight each path P with q^A, where A is the area enclosed by the path, the x-axis, and the line $x = m$, then the sum of the weights of the paths is $\begin{bmatrix} n \\ k \end{bmatrix}_q$.

In the case where the parameter q is a prime power, the Gaussian coefficients are also closely connected with the poset consisting of the subspaces of an n-dimensional vector space V over a finite field of order q. (Note that, by a theorem of Galois, a finite field has prime power order, and there is a unique field of each prime power order up to isomorphism.)

Here are some of the connections:

- The number of k-dimensional subspaces of V is $\begin{bmatrix} n \\ k \end{bmatrix}_q$.
- The Möbius function of the poset of subspaces of V (ordered by inclusion) is given by

$$\mu(A, B) = \begin{cases} (-1)^k q^{k(k-1)/2} & \text{if } A \leq B \text{ and } \dim(B/A) = k, \\ 0 & \text{otherwise.} \end{cases}$$

The value of the Möbius function can be deduced from the following result, the *q-binomial theorem*, by substituting $z = -1$.

Theorem 9. *For positive integer n and indeterminates q and z,*

$$\prod_{i=1}^{n}(1 + q^{i-1}z) = \sum_{k=0}^{n} q^{k(k-1)/2} z^k \begin{bmatrix} n \\ k \end{bmatrix}_q.$$

6.3. *Application to automata*

This previously unpublished application arose in work by Collin Bleak and the author on the outer automorphisms of the Higman–Thompson finitely presented infinite simple groups.

We require the Möbius function of the lattice of partitions of a finite set, ordered by refinement (see Exercise 9). In particular, if X is a set of cardinality n, and π_0 and π_1 are the partition with a single part and the partition into singletons respectively, then $\mu(\pi_0, \pi_1) = (-1)^{n-1}(n - 1)!$.

An *automaton* is a directed graph whose vertices are called *states* and whose edges are labelled with symbols from an alphabet of *transitions* so

that there is exactly one edge with any given label leaving any state. Imagine
the automaton as a black box with buttons on the outside: pressing one of
the buttons causes the automaton to change its internal state. We always
assume that the graph is *strongly connected*: that is, for any two states,
there is a sequence of transitions which carries one into the other.

For a positive integer k, an automaton is *k-determined* if, after the input
of k symbols, the final state depends only on the input symbols and not on
the initial state. Our question is:

Problem 6. *How many k-determined automata are there over an alphabet
of size q?*

An example of a k-determined automaton is the *de Bruijn graph*
$DB(q, k)$. Its vertices consist of all q^k words of length k over the given
alphabet. The transition y maps the state $x_1 x_2 \ldots x_k$ to the state $x_2 \ldots x_k y$.

Figure 2 shows the graph $DB(2, 3)$, with alphabet $\{0, 1\}$.

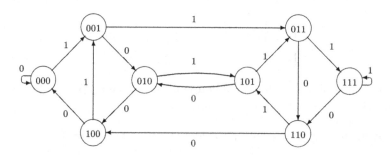

Fig. 2.   The graph $DB(2, 3)$

A *folding* of the graph $DB(q, k)$ is an equivalence relation $\equiv$ on its
vertex set with the property that, if $v \equiv w$, and y is any transition, then
the images of v and w under the transition y are also equivalent. From
a folding, we obtain an automaton as the quotient: its vertices are the
equivalence classes, and the transitions work in the obvious way. This folded
automaton is k-determined, and a moment's thought shows that any k-
determined automaton arises in this way. So we have replaced our question
with a much more concrete problem:

Problem 7. *How many foldings of $DB(q, k)$ are there?*

Let $F(q,k)$ denote the number of foldings of $\mathrm{DB}(q,k)$. We observe first:

Prop 1. $F(q,1) = B(q)$, the qth Bell number.

For the states are labelled with single transitions, and any partition of the state set is a folding.

We are going to calculate $F(q,2)$. (For larger values of k, the answer is not known.)

Let X be a set which is the disjoint union of t subsets $A_1, \ldots, A_t$, each of cardinality s. Let π be a partition of the index set $\{1, \ldots, t\}$. A partition Π of X is *compatible* with π if the sets $\bigcup\{A_i : i \in I\}$ are unions of parts of Π for any part I of π.

Let $R(s,t)$ be the number of partitions of X with the property that the only partition of $\{1, \ldots, s\}$ with which they are compatible is the partition with a single part.

Lemma 1.

$$R(s,t) = \sum_{\pi} (-1)^{|\pi|-1}(|\pi|-1)! \prod_{i=1}^{|\pi|} B(a_i s),$$

where π runs over all partitions of $\{1, \ldots, t\}$, $|\pi|$ is the number of parts of π, and a_i is the size of the ith part.

Proof. For any partition π, the number of partitions Π of X compatible with π is

$$B(a_1 s)B(a_2 s) \ldots B(a_m s),$$

where $a_1, \ldots, a_m$ are the sizes of the parts of π, and $B(n)$ is the Bell number. Now to find partitions compatible with only the trivial partition is a straightforward application of Möbius inversion over the partition lattice. $\qquad\square$

Theorem 10. *The number of foldings of $\mathrm{DB}(q,2)$ is*

$$\sum_{\pi} \prod_{i=1}^{m} R(m, a_i),$$

where π runs over all partitions of the alphabet, m is the number of parts of π, and a_i is the cardinality of the ith part for $i = 1, \ldots, m$.

Proof. Consider a folding $\equiv$ of the de Bruijn graph with word length 2 over A, where $|A| = q$. Define a partition π of A by the equivalence relation

$$a \sim b \Leftrightarrow xa \equiv yb \quad \text{for some } x, y \in A.$$

By the definition of a folding, if $a \sim b$, then $ac \equiv bc$ for all $c \in A$. So any folding class is a union of sets of the form $A_i c = \{ac : a \in A_i\}$, where A_i is a part of π, and $c \in A$. We denote this set by $A_{i,c}$. Moreover, for any folding class, the second components c of the sets $A_{i,c}$ it contains form a subset of a part B of π; so there is an induced partition of $I \times B$, where I indexes the parts of the partition. And furthermore, the definition of π shows that this partition is not compatible with any proper partition of B.

Conversely, if we have a collection of partitions of $I \times B$, for all $B \in \pi$, where the partition of $I \times B$ is not compatible with any proper partition of B, then together they form a folding of the de Bruijn graph.

So the number of foldings for given π is given by

$$\prod_{i=1}^{m} R(m, a_i),$$

where m is the number of parts of π and $a_1, \ldots, a_m$ are the sizes of the parts. The total number of foldings is given by summing this over all partitions of the alphabet. $\square$

7. Orbit Counting

This is a technique, going back to Cauchy and Frobenius, cast into the form of power series by Redfield and Pólya, for counting up to symmetry. Here is a simple example. I have a cube, and three pots of paint: red, green and blue. In how many ways can I colour the faces of the cube? Two colourings which differ only by a rotation of the cube should not be regarded as distinct.

7.1. *The orbit-counting Lemma*

We require a preliminary look at group actions. Let G be a group. An *action* of G on a set X associates a permutation π_g of X with each group element g so that the permutation associated with the group product gh is the composition of the permutations π_g and π_h. In other words, $g \mapsto \pi_g$ is a homomorphism from the group G to the set of all permutations of X. We always assume that G and X are finite.

We say that two points x, y of X lie in the same *orbit* of the action of G if there is an element $g \in G$ for which π_g maps x to y. Our job is to count the orbits. To help us, we let fix(g) denote the number of points of X which are fixed by the permutation π_g. Now the *orbit-counting Lemma* (often mistakenly called *Burnside's Lemma*) asserts that the number of orbits of the action is equal to the average value of fix(g) over all elements $g \in G$:

Theorem 11 (orbit-counting Lemma). *The number of orbits of the finite group G acting on the finite set X is*

$$\frac{1}{|G|} \sum_{g \in G} \text{fix}(g).$$

To apply this to the coloured cubes, we look at the group G of rotational symmetries of the cube. This group has 24 elements. They are classified into five types (actually *conjugacy classes*) in G as follows. For each type, we give the cycle lengths for its action on the faces of the cube. "Face axis" means an axis of rotational symmetry joining the centres of opposite faces; "edge axis" and "vertex axis" are similar.

Description	Cycles	Number
Identity	$1,1,1,1,1,1$	1
Half-turn about face axis	$1,1,2,2$	3
Quarter-turn about face axis	$1,1,4$	6
Half-turn about edge axis	$2,2,2$	6
Third-turn about vertex axis	$3,3$	8

We consider the action of G on the set of all colourings of the cube. If we have q different colours, then a colouring is fixed by a symmetry if and only if it is constant on the cycles on faces; so fix(g) $= q^{c(g)}$, where $c(g)$ is the number of cycles of g on faces. The orbit-counting Lemma gives the number of colourings up to symmetry to be

$$\tfrac{1}{24}(q^6 + 3q^4 + 12q^3 + 8q^2).$$

Setting $q = 3$, the answer to our original question is 57.

7.2. *Cycle index*

The occurrence of a polynomial in the preceding example is no surprise, as we now see.

We start with a group G acting on a set X as before. Now we take a collection F of *figures*, each with a non-negative integer *weight*. Usually the set of figures is finite, but all we require is that the number a_n of figures of weight n is finite, so we can take the *figure-counting series* $\sum a_n x^n$, the ordinary generating function for the sequence (a_n).

A *function* here will be a function from X to F; in other words, we are going to decorate every point of X with a figure. Now there is a natural action of G on the set Φ of functions, and we want to count functions up to the action of G. Each function has a total weight $\sum_{f \in F} w(f)$; and this total weight is preserved by the action of G. We want to count orbits of G on functions of total weight n. Let b_n be the total number of such orbits, and $B(n) = \sum b_n x^n$ (the *function-counting series*) its generating function.

Let $c_i(g)$ be the number of cycles of length i for the permutation π_g corresponding to g, in its action on X. We define the *cycle index* of the pair (G, X) to be the polynomial in indeterminates $s_1, s_2, \ldots$ given by

$$Z(G; s_1, s_2, \ldots) = \frac{1}{|G|} \sum_{g \in G} s_1^{c_1(g)} s_2^{c_2(g)} \ldots .$$

Now the *Cycle index theorem* asserts:

Theorem 12 (Cycle index theorem). *$B(x)$ is obtained from the cycle index $Z(G; s_1, s_2, \ldots)$ by substituting for each indeterminate s_i the power series $A(x^i)$.*

For example, our table above shows that the cycle index for the rotation group of the cube acting on faces is

$$\tfrac{1}{24}(s_1^6 + 3s_1^2 s_2^2 + 6s_1^2 s_4 + 6s_2^3 + 8s_3^2).$$

If we take three figures each of weight 0 corresponding to the three colours, then the figure-counting series is just 3, and substituting gives the same answer as before. If we take red to have weight 1 and the other two colours weight 0, the figure-counting series is $x + 2$, and the function-counting series tells us how many coloured cubes with a given number of red faces up to symmetry.

7.3. *Generalisation*

The topics of the last two sections (inclusion–exclusion and orbit counting) could be characterised by "counting with structural restrictions" and "counting up to symmetry". It is natural to ask if these two techniques can be combined.

There is no general theory, but the combination is possible in some special cases. For example, suppose we are given a graph Γ and a group G of automorphisms of Γ. Then the chromatic polynomial will count proper colourings (a structural restriction), while the orbit-counting Lemma will count all colourings up to the action of G. Now there is a polynomial associated with the pair (Γ, G), called the *orbital chromatic polynomial*, whose value at the positive integer q counts the proper colourings up to the action of G. See Cameron, Jackson & Rudd [2] for further details.

8. Species

8.1. *Labelled and unlabelled*

Problem 8. *How many graphs are there on n vertices?*

There are two interpretations of this question. In one, we take a fixed vertex set $\{1, 2, \ldots, n\}$, and count two graphs as different if their edge sets are different subsets of the set of all pairs of vertices. We say that we are counting *labelled* graphs, since each vertex comes with a label. Clearly the number in this case is the number of subsets of the set of pairs of vertices, that is, $2^{n(n-1)/2}$.

In the second interpretation, we are counting up to graph isomorphism; there is only one graph with a single edge, no matter which pair of vertices it joins. In this case we are counting *unlabelled* graphs.

It was known to graphical enumerators long ago that the appropriate generating functions to use are the exponential generating function for labelled objects, and the ordinary generating function for unlabelled objects.

From one point of view, an unlabelled object is just an orbit of the symmetric group on the underlying set of points, acting on the collection of all labelled objects. So counting unlabelled objects is an orbit-counting problem. Indeed, the best formula we have for the number of unlabelled graphs arises in precisely this way.

However, in 1981, Joyal [6] proposed a context in which this becomes clear, the theory of *species*.

8.2. *Species*

Joyal asked himself: Why can't we just take the coefficient of x^n in a formal power series to be the *set* of objects on n points, rather than just their number? A species is the thing that results from this. I will not attempt a precise definition; Joyal uses notions from category theory to do so.

Let $A(n)$ be a set of unlabelled objects on n points; we think of the corresponding species $\mathcal{A}$ as the formal power series $\sum A(n)x^n$. Now to extract information from such a gadget, we need to map each object to some numerical attribute. For a simple example, if we simply replace each object by 1, we obtain the ordinary generating function for unlabelled objects.

A profitable specialisation is to replace each object by the cycle index of its automorphism group. We denote the result (a formal power series in x where the coefficient of x^n is a polynomial in $s_1, s_2, \ldots$ with weighted degree n) by $Z(\mathcal{A})$, and call it the *cycle index* of the species.

Now it is possible to show that

- The substitution $s_i = x^i$ for all i maps $Z(\mathcal{A})$ to the ordinary generating function for the unlabelled objects in $\mathcal{A}$;
- The substitution $s_1 = x$, $s_i = 0$ for $i > 1$ maps $Z(\mathcal{A})$ to the exponential generating function for the labelled objects in $\mathcal{A}$.

One of the advantages is that, while the cycle index for an individual group can be very complicated, the sum we need to evaluate may be much simpler.

For example, consider the species $\mathcal{S}$ of sets (without any further structure): that is, $S(n)$ consists of a single n-element set. The automorphism group of $S(n)$ is the symmetric group S_n of degree n, and so

$$Z(\mathcal{S}) = \sum_{n \geq 0} Z(S_n).$$

Some calculation establishes the following remarkable identity:

Theorem 13.

$$Z(\mathcal{S}) = \prod_{i \geq 1} \exp\left(\frac{s_i}{i}\right).$$

In this case, there is a unique (labelled or unlabelled) structure of size n for any n, so the ordinary generating function for unlabelled structures is $(1 - x)^{-1}$, while the exponential generating function for labelled structures is $\exp(x)$.

Substituting x for s_1 and 0 for all the others gives $\exp(x)$, both in agreement with the general rules. On the other hand, by looking at labelled structures, we uncover the combinatorics lying behind the inverse relation between the exponential and logarithmic functions!

Substituting x^i for s_i for all i, we obtain

$$\exp\left(\sum_{i \geq 1} \frac{x^i}{i}\right) = \exp(-\log(1 - x)).$$

On the other hand, there is one labelled $\mathcal{S}$-structure on n points for any n, so the exponential generating function is

$$\sum_{n \geq 0} x^n = (1 - x)^{-1}.$$

We conclude that

$$\exp(-\log(1 - x)) = (1 - x)^{-1}.$$

Joyal developed several operations on species, mimicking those on formal power series we met earlier: addition, multiplication, differentiation and substitution. Let $\mathcal{A}$ and $\mathcal{B}$ be species.

- The *sum* $\mathcal{A} + \mathcal{B}$ is the species whose n-element structures are all the n-element structures in either of the summands.
- The *product* $\mathcal{A} \cdot \mathcal{B}$ is the species whose n-element structures are constructed as follows: choose a subset X of the n-set (of cardinality k, say); impose a k-element $\mathcal{A}$-structure on X and an $(n - k)$-element $\mathcal{B}$-structure on its complement.
- If $\mathcal{B}(0)$ is empty, then the *substitution* $\mathcal{A}[\mathcal{B}]$ is defined thus: An element of this species consists of a partition of its underlying set, an $\mathcal{A}$-structure on the set of its parts, and independently chosen $\mathcal{B}$-structures on each of the parts.

For a simple example, let $\mathcal{S}$ be the species of sets. What is $\mathcal{S} \cdot \mathcal{S}$? An object on n points is just a subset of the n-element domain; so it is the species "subset". The generating functions for unlabelled and labelled subsets are $(1 - x)^{-2}$ and $\exp(2x)$ respectively.

For another example, let $\mathcal{P}$ denote the species of permutations: each object in $\mathcal{P}(n)$ consists of a permutation of an n-set. Two permutations are isomorphic (that is, the same unlabelled structure) if they have the same cycle structure (the same number of cycles of each length); this is equivalent to conjugacy in the symmetric group. So the number of unlabelled structures in $\mathcal{P}(n)$ is the partition number $p(n)$, while the number of labelled structures is the factorial number $n!$.

Let $\mathcal{C}$ denote the species of *cycles*: each object is a set carrying a circular order, or cyclic permutation. There is a unique unlabelled object in $\mathcal{C}(n)$, and $(n-1)!$ labelled objects. (To describe a cycle, we can start at the point 1, and then the other points in order form an arbitrary permutation of $\{2,\ldots,n\}$.)

The simple fact that any permutation is a disjoint union of cycles is expressed in the language of species as

$$\mathcal{P} \sim \mathcal{S}[\mathcal{C}];$$

that is, take a partition of a set, and put a cycle on each part, to produce a permutation.

Working out what happens to the numerical power series in these cases is an interesting exercise.

8.3. *Exponential and logarithm*

As an example, we use species to give another combinatorial proof of the inverse relation between the exponential and logarithmic series.

Theorem 14. $\exp(-\log(1-x)) = (1-x)^{-1}$.

Proof. Consider the species of permutations. The exponential function for labelled permutations is $\sum n! x^n / n! = (1-x)^{-1}$.

We look more closely at the cycle decomposition. The number of cyclic permutations on an n-set is $(n-1)!$, and so the exponential generating function for these is x^n / n. It follows that the e.g.f. for arbitrary products of n-cycles is $\exp(x^n/n)$. Now an arbitrary permutation has a unique decomposition into permutations all of whose parts have the same length; so the e.g.f. for permutations is

$$\prod_n \exp\left(\frac{x^n}{n}\right) = \exp\left(\sum_n \frac{x^n}{n}\right) = \exp(-\log(1-x)).$$

The result follows. $\square$

More generally, we see that the e.g.f. for permutations all of whose cycles have lengths in a set S of positive integers is

$$\exp\left(\sum_{n \in S} \frac{x^n}{n}\right).$$

8.4. Cayley's theorem

One of the most celebrated results in enumerative combinatorics is Cayley's formula for the number of labelled trees on n vertices. A *tree* is a connected graph on n vertices without cycles.

Theorem 15 (Cayley's theorem). *The number of trees on the vertex set* $\{1,\ldots,n\}$ *is* n^{n-2}.

There are very many different proofs of Cayley's theorem. Here, I sketch two proofs which are most naturally expressed in terms of species.

First proof A *rooted tree* is a tree with a distinguished vertex called the *root*. Since there are n choices of the root, it will suffice to show that the number of rooted trees is n^{n-1}.

Let $\mathcal{R}$ be the species of rooted trees. If we remove the root from a rooted tree, we obtain a set of trees, each of which can be regarded as rooted at the vertex joined to the original root. Conversely, a set of rooted trees on $n-1$ vertices gives rise to a rooted tree on n vertices. So

$$\mathcal{R} = \mathcal{X} \cdot \mathcal{S}[\mathcal{R}],$$

where $\mathcal{S}$ is the species of sets and $\mathcal{X}$ the species of 1-element sets.

Thus, if $R(x)$ is the exponential generating function for labelled rooted trees, we have

$$R(x) = x \exp(R(x)).$$

This equation expresses the coefficients of $R(x)$ in terms of earlier coefficients. Using the *Lagrange inversion formula* (not treated here, but see the references), it can be shown that

$$R(x) = \sum_{n \geq 1} \frac{n^{n-1} x^n}{n!}.$$

Second proof This lovely proof is due to Joyal.

A *vertebrate* is a *doubly rooted tree*, that is, a tree with two distinguished vertices, which we call the *head* and the *tail*, which may or may not be equal. Since there are n^2 ways of choosing head and tail, it will suffice to show that the number of vertebrates is n^n. Note that n^n is the number of *endofunctions*, or functions from the set to itself, on a set of n vertices.

Let $\mathcal{V}$, $\mathcal{E}$, $\mathcal{L}$ and $\mathcal{P}$ be the species of vertebrates, endofunctions, linear orders and permutations. Now, in a vertebrate, there is a unique path (called the *backbone*) from the head to the tail, which is linearly ordered; and a rooted tree (possibly with just the root) is attached at each point of the backbone. So

$$\mathcal{V} \sim \mathcal{L}[\mathcal{R}],$$

where $\mathcal{R}$ is the species of rooted trees.

On the other hand, an endofunction f has a set of *periodic points* which are permuted by f. An arbitrary point arrives at a periodic point after a number of iterations of f, and the structure of the points which arrive first at the periodic point p is a tree rooted at p. So

$$\mathcal{E} \sim \mathcal{P}[\mathcal{R}].$$

Since $\mathcal{L}$ and $\mathcal{P}$ have the same number of labelled structures on n points, the same is true for $\mathcal{V}$ and $\mathcal{E}$.

9. Asymptotic Analysis

As hinted earlier, analytic techniques are useful in enumerative combinatorics in several ways (especially for estimating the order of magnitude of combinatorial functions, but also for proving identities). This is too big a subject to cover in detail here: see the works of Flajolet and Sedgewick [4] and Odlyzko [10] in the bibliography.

Instead, I conclude with a few scattered but important examples: the asymptotics of factorials, Bell numbers and numbers of unlabelled trees.

If f and g are functions on the natural numbers with $g(n) \neq 0$ for all n, we say that $f \sim g$ if the ratio $f(n)/g(n)$ tends to 1 as $n \to \infty$. Typically, f is a combinatorial counting function, and g is an analytic function whose magnitude we understand; the relation says that f and g grow at the same rate. Also, $o(1)$ denotes a function which tends to 0 as its argument goes to infinity.

9.1. *Factorials*

The most celebrated formula in asymptotic combinatorics is *Stirling's formula* for the factorials. It is easy to see that $2^{n-1} \leq n! \leq n^n$ for all n. In fact, we have a much more precise estimate:

Theorem 16 (Stirling's formula).

$$n! \sim \sqrt{2\pi n} \left(\frac{n}{e}\right)^n.$$

9.2.   *Bell numbers*

The asymptotics of the Bell numbers are rather more complicated, and can be expressed in terms of *Lambert's w-function*, the functional inverse of the function x^x (in other words, the solution to $w(x)^{w(x)} = x$). Moser and Wyman [9] proved:

Theorem 17. *The Bell numbers* $B(n)$ *satisfy*

$$B(n) \sim n! \, \frac{e^{e^r - 1}}{r^n \sqrt{2\pi r(r+1)e^r}},$$

where $r = w(n+1)$.

The import of this theorem is not very clear. It implies that, for example,

$$\frac{1}{n} \log B(n) = \log n - \log \log n - 1 + o(1),$$

(more terms in the asymptotic expansion are known).

9.3.   *Fibonacci numbers*

We saw in Section 4.1 the general form of the solution to a linear recurrence with constant coefficients. The asymptotic behaviour depends on the largest roots of the characteristic polynomial which occur in the solution, and the highest power of of n which multiply them. In particular, the Fibonacci numbers F_n satisfy

$$F_n \sim c \left(\frac{1 + \sqrt{5}}{2} \right)^n,$$

where the constant c depends on the convention for the initial values.

9.4.   *Unlabelled graphs and trees*

Usually, it is easier to count labelled objects than unlabelled. We saw that there are $2^{n(n-1)/2}$ labelled graphs, and n^{n-2} labelled trees, on n vertices. Any unlabelled object can be labelled in at most $n!$ ways, with equality if it has no non-trivial automorphisms.

For graphs, it is well-known that almost all graphs have no non-trivial automorphisms, and so the number of unlabelled graphs on n vertices is asymptotic to $2^{n(n-1)/2}/n!$ Refining the asymptotics involves considering symmetry.

For trees, however, things are more complicated. From Stirling's formula, we see that $n^{n-2}/n! \sim \frac{1}{\sqrt{2\pi}}.n^{-5/2}e^n$. However, the number of unlabelled trees is asymptotically larger than this. Otter [12] proved:

Theorem 18. *The number of unlabelled trees on n vertices is asymptotic to $cn^{-5/2}\alpha^n$, where $\alpha = 2.9557652856\ldots$ and $c = 0.534946061\ldots$.*

It follows that a random tree has exponentially many automorphisms on average.

9.5. *Binary trees*

We saw earlier that the generating function $f(x)$ for binary trees with $n+1$ leaves satisfies the recurrence

$$f(x) = x + f(x)^2.$$

The coefficients are the Catalan numbers, whose asymptotic behaviour is easily deduced from Stirling's formula:

$$C_n = \frac{1}{n+1}\binom{2n}{n} \sim \frac{1}{\sqrt{\pi}}n^{-3/2}4^n.$$

Counting these trees up to isomorphism is equivalent to removing the left–right distinction at every node. Using the cycle index $\frac{1}{2}(s_1^2 + s_2)$ for the cyclic group of order 2, we find that the functional equation for the generating function $g(x)$ of the *Wedderburn–Etherington numbers W_n* is

$$g(x) = x + \tfrac{1}{2}(g(x)^2 + g(x^2)).$$

There is no explicit solution, but the asymptotics can be worked out. We have a quadratic for $g(x)$ in terms of $g(x^2)$, with solution

$$g(x) = 1 - \sqrt{1 - 2x - g(x^2)}.$$

The asymptotics are determined by the behaviour near the closest singularity to the origin, which is a branchpoint where $g(x^2) = 1 - 2x$; note that $g(x^2)$ is analytic in the neighbourhood of this singularity. Otter found that

$$W_n \sim cn^{-3/2}\beta^n,$$

where $\beta = 2.4832535361\ldots$ and $c = 0.4026975036\ldots$. (the number of binary trees with n leaves up to isomorphism is W_{n+1}.)

The Wedderburn–Etherington numbers also have an interpretation in terms of bracketings of products, this time for powers of an element x in a commutative but non-associative algebra.

10. Further Reading

For much more on enumerative combinatorics, I recommend the books by Flajolet and Sedgewick [4] and Stanley [13] and the book-length chapter by Odlyzko [10].

Another invaluable resource for the enumerator is Neil Sloane's "On-Line Encyclopedia of Integer Sequences" [11] in which any sequence which anyone has found interesting can be looked up given the first few terms; the Encyclopedia provides references and links to related sequences.

Joyal's long paper [6] on combinatorial formal power series is in my view the best introduction to the theory of species. A book-length treatment is given by Bergeron *et al.* [1].

Like almost every area of mathematics, enumerative combinatorics throws up many interesting constants (such as those in Otter's theorem above); I recommend Finch's book [3] on *Mathematical Constants*.

Among many topics not touched on here is the connection between symmetric functions, Young tableaux, and representation theory of the symmetric group: see Stanley's book or Macdonald [8].

11. Exercises

Many of the unproved assertions in this chapter can be regarded as exercises. In addition, here are a dozen assorted exercises.

(1) Find a necessary condition on the formal power series $A_i(x)$, for $i \in \mathbb{N}$, for the infinite product $\prod\limits_i A_i(x)$ to be defined.

(2) (a) Prove that $\dbinom{-a}{k} = (-1)^k \dbinom{a+k-1}{k}$ for $k \geq 0$ and arbitrary a.

   (b) Hence verify the assertion about the Binomial Theorem for negative integer exponents after equation (1).

   (c) Show that the coefficient of x^k in $(1-x)^{-n}$ is equal to the number of selections of k things from n, with order not significant and repetition allowed.

(3) Show that the set of formal power series beginning $0 + 1x + \cdots$ forms a group under the operation of composition. (In the case where the coefficient ring is a finite field of characteristic p, this group is called the *Nottingham group* [5], and is an important example of a finitely generated pro-p-group.) What is the inverse of $x - x^2$ in this group?

(4) A *composition* of n is an ordered sequence of positive integers with sum n. (Recall that a partition of n is an unordered sequence with sum n.) Prove that the number of compositions of n is 2^{n-1} for $n \geq 1$.

(5) A permutation π of $\{1, \ldots, n\}$ is *connected* if there does not exist $i \in \{1, \ldots, n-1\}$ such that π maps the set $\{1, \ldots, i\}$ to itself. Find the coefficients of the power series

$$\left(\sum_{n \geq 0} n! x^n \right)^{-1}$$

in terms of the numbers of connected permutations of $\{1, \ldots, n\}$.

(6) (a) Show that the number $d(n)$ of derangements of $\{1, \ldots, n\}$ satisfies the recurrence

$$d(0) = 1, \qquad d(n) = n d(n-1) + (-1)^n \quad \text{for } n \geq 1.$$

(b) Show that $d(n)$ of is the nearest integer to $n! \, e^{-1}$ for $n \geq 1$.

(7) Prove Theorem 4 by applying the orbit-counting Lemma to the action of the symmetric group S_n on the set of all functions from $\{1, \ldots, n\}$ to a set of cardinality x, where x is a positive integer.

(8) (a) Show that the Boolean lattice $B(n)$ is isomorphic to the direct product of n two-element chains, and the poset of positive integers (ordered by divisibility) is isomorphic to the direct product of countably many countable chains.

(b) Show that the Möbius function of a direct product can be obtained as the product of the Möbius functions of the factors.

(c) Hence verify the formulae given in the text for the Möbius functions of the above posets.

(9) Find the Möbius function of the partition lattice.

(10) The *general linear group* $\mathrm{GL}(n, q)$, for positive integer n and prime power q, is the group of all invertible $n \times n$ matrices over the finite field of order q.

(a) Verify the following two formulae for its order:

$$\prod_{k=1}^{n} (q^n - q^{k-1}) = (-1)^n q^{n(n-1)/2} \sum_{i=0}^{n} (-1)^k q^{k(k-1)/2} \begin{bmatrix} n \\ k \end{bmatrix}_q.$$

(b) Prove that the probability that an $n \times n$ matrix over a field of order q is invertible tends to a limit $c(q)$ as $n \to \infty$ with q fixed, where $0 < c(q) < 1$. Find a convergent series expansion for $c(q)$ and use it to estimate $c(2)$ to six places of decimals.

(11) The following problem, based on the children's game "Screaming Toes", was suggested to me by Julian Gilbey.

> n people stand in a circle. Each player looks down at someone else's feet (i.e., not at their own feet). At a given signal, everyone looks up from the feet to the eyes of the person they were looking at. If two people make eye contact, they scream. What is the probability of at least one pair of people screaming?

Prove that the required probability is
$$\sum_{k=1}^{\lfloor n/2 \rfloor} \frac{(-1)^{k-1}(n)_{2k}}{(n-1)^{2k}\, 2^k\, k!}.$$

(12) (a) Show that, for fixed n and k, the limit of the q-binomial coefficient $\begin{bmatrix} n \\ k \end{bmatrix}_q$ is a polynomial in q of degree $k(n-k)$.

    (b) Find the limit of the q-binomial coefficient as q tends to a root of unity.

(13) (a) Verify the equation $\mathcal{P} \sim \mathcal{S} \cdot \mathcal{D}$, where $\mathcal{P}$, $\mathcal{S}$, and $\mathcal{D}$ are the species of permutations, sets and derangements respectively, to find the e.g.f. for derangements.

    (b) Treat similarly the equation $\mathcal{P} \sim \mathcal{S}[\mathcal{C}]$ described in the text.

    (c) Why do we require that $\mathcal{B}(0)$ should be empty in the definition of substitution $\mathcal{A}[\mathcal{B}]$?

(14) Show that, if n is even, then the number of permutations of $\{1, \ldots, n\}$ with all cycles of even length is equal to the number of permutations of this set with all cycles of odd length. Can you find a bijective proof?

(15) Use Stirling's formula to find asymptotic estimates for the binomial coefficients $\binom{n}{\alpha n}$ for fixed α with $0 < \alpha < 1$.

12.   Solution to Exercises 7, 10, and 14

Exercise 7 Let X be the set of all functions from $\{1, \ldots, n\}$ to a set of cardinality x. Then $|X| = x^n$. If g is a permutation of $\{1, \ldots, n\}$, then a function f is fixed by g if and only if it is constant on the cycles of g; so there are $x^{c(g)}$ such functions, where $c(g)$ is the number of cycles of G. The orbit-counting Lemma now shows that the number of orbits of the symmetric group S_n acting on X is
$$\frac{1}{n!} \sum_{k=1}^{n} u(n,k) x^k,$$

where $u(n, k)$ is the unsigned Stirling number of the first kind, the number of permutations with k cycles.

On the other hand, a function f can be regarded as an ordered selection of n values from a set of size x, with repetition allowed; so an orbit of S_n on functions is an unordered selection with repetition allowed. By the table of sampling rules on p. 2, the number of orbits is

$$\binom{x + n - 1}{n} = \frac{x(x + 1) \ldots (x + n - 1)}{n!}.$$

Hence

$$\sum_{k=1}^{n} u(n, k) x^k = x(x + 1) \ldots (x + n - 1).$$

Since this equation is true for any positive integer x, it is a polynomial identity. Now replacing x by $-x$, multiplying by $(-1)^n$, and using the fact that $s(n, k) = (-1)^{n-k} u(n, k)$, we have the result.

Exercise 10   (a) Choose a fixed basis B for the n-dimensional vector space V over a field of order q. Now any invertible linear map takes B to a basis, and for any basis B' there is a unique invertible map carrying B to B'. So the order of the general linear group is equal to the number of choices of a basis.

- There are $q^n - 1$ choices for the first vector of a basis, since there are q^n vectors in the space and any except the zero vector can be chosen.
- The second basis vector can be any vector which is not a multiple of the first; so there are $q^n - q$ choices.
- The third basis vector can be any vector which is not a linear combination of the first two; so there are $q^n - q^2$ choices.
- And so on ....

Multiplying the numbers of choices gives the formula on the left.

Now this can be rewritten as

$$|\mathrm{GL}(n, q)| = (-1)^n q^{n(n-1)/2} \prod_{i=1}^{n} (1 - q^i)$$

and the right-hand side is obtained by substituting $z = -q$ in the q-binomial theorem.

(b) The total number of $n \times n$ matrices is q^{n^2}, so the probability that a random matrix is invertible is

$$c_n(q) = \prod_{i=1}^{n}(1 - q^{-i}).$$

As $n \to \infty$, we have

$$c_n(q) \to c(q) = \prod_{i \geq 1}(1 - q^{-i}).$$

(This infinite product *converges*, which means that the limit is strictly between 0 and 1.) According to Euler's Pentagonal Numbers Theorem, we have

$$c(q) = \sum_{k \in \mathbb{Z}}(-1)^k q^{-k(3k-1)/2} = 1 - q^{-1} - q^{-2} + q^{-5} + q^{-7} - q^{-12} - \cdots.$$

So, for example, the limiting probability that a large random matrix over the integers mod 2 is invertible is $0.288788\ldots$. (Observe that the series converges much faster than the infinite product.)

Exercise 14 The e.g.f. for permutations on sets of even size is $\sum_{n \text{ even}} x^n = (1 - x^2)^{-1}$. Let $E(x)$ and $O(x)$ be the e.g.f.s for permutations (on sets of even size) with all cycles even, resp. odd. Since every permutation can be decomposed uniquely into permutations with even and odd cycles respectively, we see that $E(x) \cdot O(x) = (1 - x^2)^{-1}$.

By the remark after Theorem 14,

$$E(x) = \exp\left(\sum_{n \text{ even}} \frac{x^n}{n}\right) = \exp(-\log(1 - x^2)/2) = (1 - x^2)^{-1/2}.$$

Hence, $E(x) = O(x)$.

For a bijective proof, see Lewis and Norton [7].

References

[1] F. Bergeron, G. Labelle and P. Leroux, *Combinatorial Species and Tree-like Structures*, Encyclopedia of Mathematics and its Applications **67**, Cambridge University Press, Cambridge, 1998.
[2] P. J. Cameron, B. Jackson and J. Rudd, Orbit-counting polynomials for graphs and codes, *Discrete Math.* **308** (2008), 920–930.
[3] S. R. Finch, *Mathematical Constants*, Encyclopedia of Mathematics and its Applications **94**, Cambridge University Press, Cambridge, 2003.

[4] P. Flajolet and R. Sedgewick, *Analytic Combinatorics*, Cambridge University Press, Cambridge, 2009.

[5] D. L. Johnson, The group of formal power series under substitution, *J. Austral. Math. Soc.* (A) **45** (1988), 296–302.

[6] A. Joyal, Une théorie combinatoire des séries formelles, *Advances in Mathematics* **42** (1981), 1–82.

[7] R. P. Lewis and S. P. Norton, On a problem raised by P. J. Cameron, *Discrete Math.* **138** (1995), 315–318.

[8] I. G. Macdonald, *Symmetric functions and Hall polynomials*, Oxford University Press, Oxford, 1995.

[9] L. Moser and M. Wyman, An asymptotic formula for the Bell numbers, *Trans. Roy. Soc. Canada* **49** (1955), 49–54.

[10] A. M. Odlyzko, Asymptotic enumeration methods, pp. 1063–1230 in *Handbook of Combinatorics* (eds. R. L. Graham, M. Grötschel and L. Lovász), Elsevier, Amsterdam, 1995.

[11] *The On-Line Encyclopedia of Integer Sequences*, https://oeis.org/.

[12] R. Otter, The number of trees, *Ann. Math.* **49** (1948), 583–599.

[13] R. P. Stanley, *Enumerative Combinatorics* (2 volumes), Cambridge University Press, Cambridge, 1999, 2012.

Chapter 2

Introduction to the Finite Simple Groups

Robert A. Wilson

School of Mathematical Sciences,
Queen Mary University of London, London E1 4NS, UK
r.a.wilson@qmul.ac.uk

The complete determination of all the finite simple groups is one of the major achievements of 20th century mathematics. Yet, even the statement of this theorem, let alone its proof, appears impenetrable to anyone except a professional group theorist. It is my aim in this chapter to demystify the statement of the theorem, by giving a practical introduction to the various types of finite simple groups.

1. Overview

Simple groups A subgroup H of a group G is *normal* if the left and right cosets are equal, $Hg = gH$ for all $g \in G$. A group G is *simple* if it has exactly two normal subgroups, 1 and G.

The Abelian simple groups are exactly the cyclic groups of prime order, C_p. The non-Abelian simple groups are much harder to classify: 50 years of hard work by many people, c. 1955–2005, led to *CFSG*: the Classification Theorem for Finite Simple Groups:

Theorem 1 (CFSG). *Every non-Abelian finite simple group is one of the following:*

- *An alternating group A_n, $n \geq 5$: the set of even permutations of n points;*
- *A classical group over a finite field: six families (linear, unitary, symplectic, and three families of orthogonal groups);*
- *An exceptional group of Lie type: ten families;*
- *26 sporadic simple groups, ranging in size from M_{11} of order 7920 to the Monster of order nearly 10^{54}.*

Classical groups   The six families of classical finite simple groups are all essentially matrix groups over finite fields:

- The *projective special linear groups* $PSL_n(q)$;
- The *projective special unitary group* $PSU_n(q)$;
- The *projective symplectic groups* $PSp_{2n}(q)$;
- Three families of *orthogonal groups*,
  - $P\Omega_{2n+1}(q)$;
  - $P\Omega_{2n}^+(q)$;
  - $P\Omega_{2n}^-(q)$.

The first of these families, the linear groups, is the subject of Sections 4 and 5. All the other groups are obtained by fixing various types of *forms*, which are the subject of Section 6. The groups themselves are constructed in Section 7.

Exceptional groups   The ten families of exceptional groups of Lie type are $G_2(q)$, $F_4(q)$, $E_6(q)$, $E_7(q)$, $E_8(q)$, $^2E_6(q)$, $^3D_4(q)$, $^2B_2(q)$, $^2G_2(q)$, $^2F_4(q)$. They are all derived in some way from *Lie algebras*. We shall focus on $G_2(q)$ in Section 9 and $F_4(q)$ and $E_6(q)$ in Section 10, after a brief look at the general Lie theory in section 8.

Sporadic groups   The 26 sporadic simple groups may be roughly divided into five types:

- The five *Mathieu groups* M_{11}, M_{12}, M_{22}, M_{23}, M_{24}:
  - permutation groups on 11, ..., 24 points.
- The seven *Leech lattice groups* Co_1, Co_2, Co_3, McL, HS, Suz, J_2:
  - (real) matrix groups in dimension at most 24.
- The three *Fischer groups* Fi_{22}, Fi_{23}, Fi'_{24}:
  - automorphism groups of rank 3 graphs.
- The five *Monstrous groups* $\mathbb{M}$, $\mathbb{B}$, Th, HN, He:
  - centralisers in the Monster of elements of order 1, 2, 3, 5, 7.
- The six *pariahs* J_1, J_3, J_4, ON, Ly, Ru:
  - oddments which have little to do with each other.

Our study will be restricted to the first two families, the Mathieu groups in Section 11 and the Leech lattice groups in Section 12.

2. Alternating Groups

Permutations   A *permutation* on a set Ω is a bijection from Ω to itself. The set of permutations on Ω forms a group, called the *symmetric group* on Ω.

A *transposition* is a permutation which swaps two points and fixes all the rest. Every permutation can be written as a product of transpositions. The identity element cannot be written as the product of an odd number of transpositions. Hence, no element can be written both as an even product and an odd product.

Even permutations   A permutation is *even* if it can be written as a product of an even number of transpositions, and *odd* otherwise. The even permutations form a subgroup called the *alternating group*, and the odd permutations form a coset of this subgroup. In particular, the alternating group has index 2 in the symmetric group. So if Ω has n points, the symmetric group S_n has order $n!$, and the alternating group has order $n!/2$.

Transitivity   Write a^π for the image of $a \in \Omega$ under the permutation π. The *orbit* of $a \in \Omega$ under the group H is $\{a^\pi \mid \pi \in H\}$. The orbits under H form a partition of Ω.

If there is only one orbit (Ω itself), then H is *transitive*. For $k \geq 1$, a group H is *k-transitive* if for every list of k distinct elements $a_1, \ldots, a_k \in \Omega$ and every list of k distinct elements $b_1, \ldots, b_k \in \Omega$, there is a permutation $\pi \in H$ with $a_i{}^\pi = b_i$ for all i.

Primitivity   A *block system* for H is a partition of Ω preserved by H. The partitions $\{\Omega\}$ and $\{\{a\} \mid a \in \Omega\}$ are *trivial* block systems. If H preserves a non-trivial block system (called a *system of imprimitivity*), then H is called *imprimitive*. Otherwise H is primitive. If H is primitive, then H is transitive, because the orbits of an intransitive group form a system of imprimitivity. If H is 2-transitive, then H is primitive, because if there were a non-trivial block system then all pairs would have to lie in the same block.

Group actions   Suppose G is a subgroup of S_n acting on $\Omega = \{1, 2, \ldots, n\}$. The *stabilizer* of $a \in \Omega$ in G is $H := \{g \in G \mid a^g = a\}$. The set $\{g \in G \mid a^g = b\}$ is equal to the coset Hx, where x is any element with $a^x = b$. In other words $a^x \mapsto Hx$ is a bijection between Ω and the set of right cosets of H. Hence the *orbit-stabilizer theorem*: $|H| \cdot |\Omega| = |G|$.

Conversely, the action of G on Ω is the same as (equivalent to) the action on cosets of H given by $g : Hx \mapsto Hxg$.

Maximal subgroups This equivalence gives a very useful correspondence between *transitive group actions* and *subgroups*, in which *primitive* group actions correspond to *maximal* subgroups:

Proof. The block of imprimitivity containing a, say B, corresponds to the cosets Hx such that $a^x \in B$. The union of these cosets is a subgroup K with $H < K < G$, so H is not maximal. And conversely.

Conjugacy classes Every element in S_n can be written as a product of disjoint cycles. Conjugation by $g \in S_n$ is the map $x \mapsto g^{-1}xg$. It maps a cycle $(a_1, \ldots, a_k)$ to $(a_1{}^g, \ldots, a_k{}^g)$. Hence two elements of S_n are *conjugate* if and only if they have the same *cycle type*.

Conjugacy in A_n is a little more subtle: if there is a cycle of even length, or two cycles of the same odd length, then we get the same answer. But if the cycles have distinct odd lengths then the conjugacy class in S_n splits into two classes of equal size in A_n. This is because these are exactly the elements of A_n which do not commute with any odd permutation.

Simplicity of A_5 There are just five conjugacy classes in A_5, consisting of, respectively:

- One identity element;
- 15 elements of shape $(a,b)(c,d)$;
- 20 elements of shape (a,b,c);
- 12 elements of shape (a,b,c,d,e): two such conjugacy classes.

No proper non-trivial union of conjugacy classes, containing the identity element, has size dividing 60, so there is no proper non-trivial normal subgroup.

Simplicity of A_n Assume N is a normal subgroup of A_n and $n \geq 6$. Then $N \cap A_{n-1}$ is normal in A_{n-1}, so by induction is either 1 or A_{n-1}. In the first case, N has at most n elements, but there is no conjugacy class small enough to be in N. (This is slightly awkward to prove rigorously.) In the second case, N contains a 3-cycle, so contains all 3-cycles, so is A_n.

3. Subgroups of Symmetric Groups

We work in S_n rather than A_n (as it is easier), and consider only maximal subgroups. The O'Nan–Scott Theorem says that every maximal subgroup of A_n or S_n is of one of six types. Five of these are explicitly defined subgroups, while the last is just a general description of a class of rather rare small maximal subgroups. Before stating the theorem, we first describe these types in some detail.

Intransitive subgroups If a subgroup has more than two orbits, it cannot be maximal. If a subgroup has two orbits, of lengths k and $n - k$, then it is contained in $S_k \times S_{n-k}$. This is maximal if $k \neq n - k$.

If $k = n - k$, we can adjoin an element swapping the two orbits, giving a larger group $(S_k \times S_k) \cdot 2$ which is maximal.

Therefore, the *intransitive maximal subgroups* of S_n are, up to conjugacy, $S_k \times S_{n-k}$ for $1 \leq k < n/2$.

Transitive imprimitive subgroups If $n = km$, then you can split Ω into k disjoint subsets of size m. The stabilizer of this partition contains $S_m \times S_m \times \cdots \times S_m$, the direct product of k copies of S_m. It also contains S_k permuting the k blocks. Together, these form the *wreath product* of S_m with S_k, written $S_m \wr S_k$. These subgroups are again maximal in S_n.

Primitive wreath products If $n = m^k$, we can label the n points of Ω by k-tuples $(a_1, \ldots, a_k)$ of elements a_i from a set A of size m. Then $S_m \times S_m \times \cdots \times S_m$ can act on this set by getting each copy of S_m to act on one of the k coordinates. Also, S_k can act by permuting the k coordinates. This gives an action of $S_m \wr S_k$ on the set of m^k points. This is called the *product action* to distinguish it from the *imprimitive action* we have just seen. This group is not maximal if $m \leq 4$. If $m \geq 5$ it is maximal in either A_n or S_n, depending on whether k^{m-1} is or is not divisible by 4.

Affine groups If $n = p^d$, where p is prime, then we can label the n points of Ω by the vectors of a d-dimensional vector space over $\mathbb{Z}/p\mathbb{Z}$. The *translations* $x \mapsto x + v$ act on this vector space. The *linear transformations* $x \mapsto xM$ (where M is an invertible matrix) also act. These generate a group $AGL_d(p)$, the *affine general linear group*. These groups are not always maximal in S_n: for example, they are often contained in A_n.

Subgroups of diagonal type   These are harder to describe. Let T be a non-Abelian simple group, and let H be the wreath product $T \wr S_m$ for some $m \geq 2$. This contains a "diagonal" subgroup $D \cong T$ consisting of all the "diagonal" elements $(t, t, \ldots, t) \in T \times T \times \cdots \times T$.

Indeed, H contains a subgroup $D \times S_m$ of index $|T|^{m-1}$. Let H act on the $n = |T|^{m-1}$ cosets of this subgroup. Then H is nearly maximal in S_n: we just need to adjoin the automorphisms of T, acting the same way on all the m copies of T.

Almost simple groups   A group G is *almost simple* if there is a simple group T such that $T \leq G \leq \mathrm{Aut}T$. If M is a maximal subgroup of G, then G acts primitively on the $|G|/|M|$ cosets of M. Hence, G is a subgroup of S_n, where $n = |G|/|M|$.

Often, such a group G is maximal in S_n or A_n. For a reasonable value of n, these are straightforward to classify. But classifying these groups G for all n is a hopeless task.

Theorem 2 (O'Nan–Scott). *If H is any proper subgroup of S_n other than A_n, then H is a subgroup of (at least) one of the following:*

- *(intransitive) $S_k \times S_{n-k}$, for $k < n/2$;*
- *(transitive imprimitive) $S_k \wr S_m$, for $n = km$, $1 < k < n$;*
- *(product action) $S_k \wr S_m$, for $n = k^m$, $k \geq 5$;*
- *(affine) $AGL_d(p)$, for $n = p^d$, p prime;*
- *(diagonal) $T^m.(\mathrm{Out}(T) \times S_m)$, where T is non-Abelian simple, and $n = |T|^{m-1}$;*
- *(almost simple) an almost simple group G acting on the n cosets of a maximal subgroup M of G.*

4.   Linear Groups

Finite fields   A *field* is a set F with all the usual arithmetical operations and rules.

- $\{F, +, -, 0\}$ is an Abelian group;
- $\{F^*, ., /, 1\}$ is an Abelian group, where $F^* = F \backslash \{0\}$;
- $x(y + z) = xy + xz$.

Example: if p is a prime, then the integers modulo p form a field $\mathbb{F}_p = \mathbb{Z}/p\mathbb{Z}$.

In any finite field F, the subfield F_0 generated by 1 has prime order, p. This prime is called the *characteristic* of F. Moreover, F is a vector space, of dimension d, over F_0, so has p^d elements. In fact, there is exactly one field of each such order $q = p^d$. Denote this field $\mathbb{F}_q$. To make such a field, pick an irreducible polynomial f of degree d, and construct the quotient $F_0[x]/(f)$ of the polynomial ring.

Example: $p = 2$, with $f(x) = x^2 + x + 1$, gives a field of order 4 as $\mathbb{F}_4 = \{0, 1, \omega, \overline{\omega}\}$ with $\omega^2 = \overline{\omega}$ and $\omega + \overline{\omega} = 1$.

The general linear group $GL_n(q)$ is the group of all invertible $n \times n$ matrices with entries in the field $F = \mathbb{F}_q$ of order q. The scalar matrices form a normal subgroup Z of order $q-1$. The *projective general linear group* $PGL_n(q) = GL_n(q)/Z$.

The determinant map $\det : GL_n(q) \to F^*$ is a group homomorphism. Its kernel is the *special linear group* $SL_n(q)$. The *projective special linear group* $PSL_n(q) = SL_n(q)/(Z \cap SL_n(q))$.

The orders of the linear groups How many invertible matrices are there? Choose each row to be linearly independent of the previous rows. The first i rows span an i-dimensional space, which has q^i vectors. Therefore, there are $q^n - q^i$ choices for the $(i + 1)$th row. Hence,

- $|GL_n(q)| = (q^n - 1)(q^n - q) \ldots (q^n - q^{n-1})$.
- $|SL_n(q)| = |PGL_n(q)| = |GL_n(q)|/(q - 1)$.
- $|PSL_n(q)| = |SL_n(q)|/\gcd(n, q - 1)$.

An example: $GL_2(2)$ Take the field $F = \mathbb{F}_2 = \mathbb{Z}/2\mathbb{Z} = \{0, 1\}$ with $1 + 1 = 0$. The two-dimensional vector space F^2 has four vectors, $(0, 0)$, $(0, 1)$, $(1, 0)$, $(1, 1)$. The first row can be any of the three non-zero vectors. The second row can be any of the remaining two non-zero vectors. Hence, $|GL_2(2)| = 6$.

Now $GL_2(2)$ acts on the vector space by permuting the three non-zero vectors in all possible ways. Hence, $GL_2(2) \cong S_3$.

More examples

- $PGL_2(3) \cong S_4$, permuting the four one-dimensional subspaces;
- $PSL_2(3) \cong A_4$;
- $PSL_2(4) \cong A_5$, permuting the five one-dimensional subspaces;
- $PSL_2(5) \cong A_5$, and $PGL_2(5) \cong S_5$.

In fact, $PSL_n(q)$ is a simple group except for the cases $PSL_2(2) \cong S_3$ and $PSL_2(3) \cong A_4$. The easiest way to prove simplicity of $PSL_n(q)$ is to use:

Theorem 3 (Iwasawa's lemma). *If G is a finite perfect group acting faithfully and primitively on a set Ω, and the point stabilizer H has a normal Abelian subgroup A whose conjugates generate G, then G is simple.*

Proof. Otherwise, choose a normal subgroup K with $1 < K < G$. Choose a point stabilizer H with $K \not\leq H$. Hence $G = HK$ since H is maximal.

Any $g \in G$ can be written $g = hk$. Any conjugate of A is $g^{-1}Ag = k^{-1}h^{-1}Ahk = k^{-1}Ak \leq AK$. Therefore $G = AK$.

Now $G/K = AK/K \cong A/(A \cap K)$ is Abelian. But G is perfect, that is, $G = G'$, so has no non-trivial Abelian quotients. Contradiction.

Simplicity of $PSL_n(q)$ Let $PSL_n(q)$ act on the one-dimensional subspaces of F^n. This action is 2-transitive, so primitive. Write this action as a right action of matrices on row vectors.

The stabilizer of the point $\langle (1, 0, 0, \ldots, 0) \rangle$ consists (modulo scalars) of matrices $\begin{pmatrix} \lambda & 0 \\ v & M \end{pmatrix}$. It has a normal Abelian subgroup consisting of matrices $\begin{pmatrix} 1 & 0 \\ v & I_{n-1} \end{pmatrix}$.

These matrices encode elementary row operations, and it is an elementary theorem of linear algebra that every matrix of determinant 1 is a product of such matrices. To prove that $PSL_n(q)$ is perfect, it suffices to prove that these matrices (*transvections*) are commutators. If $n \geq 3$, observe that

$$\left[\begin{pmatrix} 1 & 0 & 0 \\ 1 & 1 & 0 \\ 0 & 0 & 1 \end{pmatrix}, \begin{pmatrix} 1 & 0 & 0 \\ 0 & 1 & 0 \\ 0 & x & 1 \end{pmatrix} \right] = \begin{pmatrix} 1 & 0 & 0 \\ 0 & 1 & 0 \\ -x & 0 & 1 \end{pmatrix}.$$

If $q \geq 4$, then $\mathbb{F}_q$ has an element x with $x^3 \neq x$, so

$$\left[\begin{pmatrix} 1 & 0 \\ y & 1 \end{pmatrix}, \begin{pmatrix} x & 0 \\ 0 & x^{-1} \end{pmatrix} \right] = \begin{pmatrix} 1 & 0 \\ y(x^2 - 1) & 1 \end{pmatrix}.$$

Hence, by Iwasawa's lemma, $PSL_n(q)$ is simple whenever $n \geq 3$ or $q \geq 4$.

5. Subgroups of General Linear Groups

Subspace stabilizers The stabilizer of a subspace of dimension k looks like this:

$$\begin{pmatrix} GL_k(q) & 0 \\ q^{k(n-k)} & GL_{n-k}(q) \end{pmatrix}.$$

The subgroup of matrices of shape $\begin{pmatrix} I_k & 0 \\ A & I_{n-k} \end{pmatrix}$ is a normal elementary Abelian subgroup, of order $q^{k(n-k)}$. The quotient by this subgroup is isomorphic to $GL_k(q) \times GL_{n-k}(q)$.

Imprimitive subgroups Suppose $V = V_1 \oplus \cdots \oplus V_m$ is a direct sum of m subspaces each of dimension k. The stabiliser of this decomposition of the vector space has a normal subgroup $GL_k(q) \times \cdots \times GL_k(q)$ acting on $V_1, \ldots, V_m$ separately. There is also a subgroup S_m permuting these m subspaces. Together, these generate a wreath product $GL_k(q) \wr S_m$.

Tensor products If $A = (a_{ij}) \in GL_k(q)$ and $B = (b_{ij}) \in GL_m(q)$ then the following matrix is in $GL_{km}(q)$:

$$A \otimes B = \begin{pmatrix} a_{11}B & a_{12}B & \cdots & a_{1k}B \\ a_{21}B & a_{22}B & \cdots & a_{2k}B \\ \vdots & & & \\ a_{k1}B & a_{k2}B & \cdots & a_{kk}B \end{pmatrix}.$$

If we multiply A by a scalar λ, and B by the inverse λ^{-1}, then this matrix does not change. Factoring out by the scalars, therefore, we have

$$PGL_k(q) \times PGL_m(q) < PGL_{km}(q).$$

Wreathed tensor products Repeating this construction with m copies of $GL_k(q)$, we get

$$PGL_k(q) \times \cdots \times PGL_k(q) < PGL_{k^m}(q).$$

We can also permute the m copies of $PGL_k(q)$ with a copy of S_m. Together, these give a wreath product

$$PGL_k(q) \wr S_m < PGL_{k^m}(q).$$

Extraspecial groups   Suppose r is an odd prime, and α is an element of order r in the field F (so r is a divisor of $|F| - 1$). Let R be the group generated by the $r \times r$ matrices

$$\begin{pmatrix} \alpha & 0 & 0 & \cdots & & 0 \\ 0 & \alpha^2 & 0 & \cdots & & 0 \\ \vdots & & & & & \\ 0 & 0 & & \cdots & & 0 \\ 0 & 0 & & \cdots & \alpha^{r-1} \end{pmatrix} \quad \text{and} \quad \begin{pmatrix} 0 & 1 & 0 & \cdots & 0 \\ 0 & 0 & 1 & & 0 \\ \vdots & & & & \\ 0 & 0 & 0 & \cdots & 1 \\ 1 & 0 & 0 & \cdots & 0 \end{pmatrix}.$$

Then R is non-Abelian of order r^3. Taking the tensor product of k copies of R gives an *extraspecial group* of order r^{1+2k} acting on a space of dimension r^k. This representation is irreducible over any field of characteristic $p \neq r$, and extends to a group $r^{1+2k}{:}Sp_{2k}(r)$.

A slightly different construction is required for the prime 2. Both D_8 and Q_8 have two-dimensional representations. Tensoring them together gives extraspecial groups of order 2^{1+2k}, with representations of degree 2^k. In fact, $D_8 \otimes D_8 = Q_8 \otimes Q_8$, so there are just two extraspecial groups of each order 2^{1+2k}. In this case, we can extend by an orthogonal group rather than a symplectic group.

If the field contains elements of order 4, we can adjoin these scalars to make a bigger group which contains both extraspecial groups. This allows us to extend by the full symplectic group.

Almost quasi-simple groups   A group G is *quasi-simple* if $G = G'$ and $G/Z(G)$ is simple. Example: $SL_n(q)$ is quasi-simple except for $SL_2(2)$ and $SL_2(3)$. A group G is *almost quasi-simple* (for us — this is not entirely standard terminology) if G/Z is almost simple, where Z is a group of scalar matrices.

Theorem 4 (Aschbacher–Dynkin). *Every subgroup of $GL_n(q)$ which does not contain $SL_n(q)$ is contained in the stabiliser of one of the following:*

- *A subspace of dimension k;*
- *A direct sum of k subspaces of dimension m, where $n = km$;*
- *A tensor product $F^k \otimes F^m$, where $n = km$;*
- *A tensor product of m copies of F^k, where $n = k^m$;*
- *An extraspecial group r^{1+2k}, where $n = r^k$;*
- *An almost quasi-simple subgroup, acting irreducibly.*

6. Forms

Bilinear forms A *bilinear form* on a vector space V is a map $B :$ $V \times V \to F$ satisfying

$$B(\lambda u + v, w) = \lambda B(u, w) + B(v, w),$$
$$B(u, \lambda v + w) = \lambda B(u, v) + B(u, w).$$

It is

- *Symmetric if $B(u, v) = B(v, u)$,*
- *Skew-symmetric if $B(u, v) = -B(v, u)$,*
- *Alternating if $B(v, v) = 0$.*

An alternating bilinear form is always skew-symmetric, but the converse is true if and only if the characteristic is not 2.

Quadratic forms A *quadratic form* is a map $Q : V \to F$ satisfying

$$Q(\lambda u + v) = \lambda^2 Q(u) + \lambda B(u, v) + Q(v)$$

where B is the *associated bilinear form*. The quadratic form can be recovered from the bilinear form, as $Q(v) = \frac{1}{2}B(v, v)$, if and only if the characteristic is not 2.

In characteristic 2, the associated bilinear form is alternating, since

$$0 = Q(0) = Q(v + v) = 2Q(v) + B(v, v) = B(v, v).$$

Sesquilinear forms Let F be the field of order q^2, and let $^-$ denote the field automorphism $x \mapsto x^q$. Then $B : V \times V \to F$ is *sesquilinear* if

$$B(\lambda u + v, w) = \lambda B(u, w) + B(v, w)$$

and *conjugate-symmetric* if

$$B(w, v) = \overline{B(v, w)}.$$

If both these conditions hold, then also

$$B(u, \lambda v + w) = \overline{\lambda} B(u, v) + B(u, w).$$

Properties of forms Two vectors u, v are *perpendicular*, written $u \perp v$, if $B(u,v) = 0$. We define $S^\perp = \{v \in V \mid x \perp v \text{ for all } x \in S\}$, and say that v is *isotropic* if $B(v,v) = 0$ (or $Q(v) = 0$).

The *radical* rad(B) of B is $V^\perp$. Also, B is *non-singular* if rad$(B) = 0$, and *singular* otherwise. In this situation, B is sometimes called an *inner product*. Similarly, the radical of Q is the subspace of isotropic vectors in the radical of the associated B.

A subspace is *non-singular* if the form restricted to the subspace is non-singular. A subspace is *totally isotropic* if the form restricted to the subspace is identically zero.

Isometries and similarities An *isometry* of B (a form on V) is a linear map $\phi : V \to V$ which preserves the form, in the sense $B(u^\phi, v^\phi) = B(u,v)$. Similarly, an isometry of Q is a linear map ϕ which satisfies $Q(v^\phi) = Q(v)$.

A *similarity* allows changes of scale: that is

$$B(u^\phi, v^\phi) = \lambda_\phi B(u,v)$$

or

$$Q(v^\phi) = \lambda_\phi Q(v).$$

Classification of alternating bilinear forms If we can find vectors u,v such that $B(u,v) = \lambda \neq 0$, then take our first two basis vectors to be u and $\lambda^{-1}v$, so that the form has matrix

$$\begin{pmatrix} 0 & 1 \\ -1 & 0 \end{pmatrix}.$$

Now restrict to $\{u,v\}^\perp$ and continue. When there are no such vectors left, the form is identically zero.

Notice that the rank of B is always even. Up to change of basis, there is a unique non-singular form.

Classification of sesquilinear forms If there is a vector v with $B(v,v) = \lambda \neq 0$, then $\lambda = \overline{\lambda}$ which implies that there exists $\mu \in F$ with $\mu\overline{\mu} = \mu^{q+1} = \lambda$. Therefore, $v' = \mu^{-1}v$ satisfies $B(v',v') = 1$. Now restrict to $v^\perp$ and continue.

If there is no such v, then we can easily show that the form is identically zero. Again, there is a unique non-singular form, up to change of basis.

Classification of symmetric bilinear forms in odd characteristic   We can diagonalise the form as in the sesquilinear case, but adjusting the scalars requires more care.

If $B(v,v) = \lambda$ is a square, $\lambda = \mu^2$, then we can replace v by $v' = \mu^{-1}v$ and get $B(v',v') = 1$. But if $B(v,v)$ is not a square, the best we can do is adjust it to be equal to our favourite non-square α, say.

Now we can replace two copies of α by two copies of 1, by picking λ and μ such that $\lambda^2 + \mu^2 = \alpha^{-1}$, and changing basis via $x' = \lambda x + \mu y$ and $y' = \mu x - \lambda y$. It follows that there are exactly two non-singular forms, up to change of basis.

Classification of quadratic forms in characteristic 2   Again we find that there are exactly two non-singular forms, up to change of basis. The first one has matrix equal to the identity matrix, and is called of *plus type*.

The second one has a 2×2 block $\begin{pmatrix} 1 & 1 \\ 0 & \mu \end{pmatrix}$ where $x^2 + x + \mu$ is irreducible over $\mathbb{F}_q$, and is called of *minus type*.

Theorem 5 (Witt's Lemma).   *If (V, B) and (W, C) are isometric spaces, with B and C non-singular, and either*

- *Alternating bilinear, or*
- *Conjugate-symmetric sesquilinear, or*
- *Symmetric bilinear in odd characteristic*

then any isometry between a subspace X of V and a subspace Y of W extends to an isometry of V with W.

7.   Definitions of the Classical Groups

Symplectic groups   The *symplectic group* $Sp_{2n}(q)$ is the isometry group of a non-singular alternating bilinear form on $V = \mathbb{F}_q^{2n}$.

To calculate its order, count the number of ways of choosing a standard basis. Pick the first vector in $q^{2n} - 1$ ways. Of the $q^{2n} - q$ vectors which are linearly independent of the first, $q^{2n-1} - q$ are orthogonal to it, and q^{2n-1} have each non-zero inner product with it. So there are q^{2n-1} choices for the second vector.

By induction on n, the order of $Sp_{2n}(q)$ is

$$\prod_{i=1}^{n}(q^{2i} - 1)q^{2i-1} = q^{n^2}\prod_{i=1}^{n}(q^{2i} - 1).$$

Structure of symplectic groups   It is easy to see that the only
scalars in $Sp_{2n}(q)$ are ± 1. Every element in $Sp_{2n}(q)$ has determinant 1.
(This is unfortunately not obvious.)

We show $Sp_2(q) \cong SL_2(q)$, by direct calculation: $\begin{pmatrix} a & b \\ c & d \end{pmatrix}$ preserves
the standard symplectic form if and only if $B((a,b),(c,d)) = 1$, that is
$ad - bc = 1$.

In fact, $Sp_4(2) \cong S_6$. All other projective symplectic groups $PSp_{2n}(q)$,
for $n \geq 2$, are simple. (Proof using transvections and Iwasawa's Lemma as
for $PSL_n(q)$.)

Unitary groups   The *(general) unitary group* $(G)U_n(q)$ is the isometry group of a non-singular conjugate-symmetric sesquilinear form on V of
dimension n over F_{q^2}.

It is not quite so easy to calculate the order this time. Induction on n
gives the number of vectors of norm 1 as

$$q^{n-1}(q^n - (-1)^n).$$

Then another induction on n gives the order of the group as

$$\prod_{i=1}^{n} q^{i-1}(q^i - (-1)^i) = q^{\frac{n(n-1)}{2}} \prod_{i=1}^{n}(q^i - (-1)^i).$$

Structure of unitary groups $M \in U_n(q)$ iff $M\overline{M}^T = I_n$. In partic-
ular, if $\det(M) = \lambda$ then $\lambda\bar{\lambda} = 1$, and there are $q+1$ possibilities for λ. The
special unitary group $SU_n(q)$ is the subgroup of matrices of determinant 1,
and is a normal subgroup of index $q + 1$.

The scalars in $GU_n(q)$ are those satisfying $\lambda.\bar{\lambda} = 1$, so form a normal
subgroup of order $q + 1$. The scalars in $SU_n(q)$ form a group of order
$(n, q+1)$.

We have $PSU_2(q) \cong PSL_2(q)$. In general, $PSU_n(q)$ is simple, except
that $PSU_3(2)$ has order $72 = 2^3 \cdot 3^2$ so is not simple (e.g., by Burnside's
$p^a q^b$-theorem). Indeed $PSU_3(2) \cong 3^2{:}Q_8$ and $PGU_3(2) \cong 3^2{:}SL_2(3)$.

Orthogonal groups, odd characteristic   The orthogonal groups
are the isometry groups of non-singular symmetric bilinear forms. Since
there are two types of forms, there are two types of groups. But in odd
dimensions, the two types of forms are scalar multiples of each other, so
the two groups are the same.

In even dimensions, $2n$ say, the form has *plus type* if there is a totally isotropic subspace of dimension n. This is *not the same as having an orthonormal basis*. The other forms have *minus type*, and their maximal totally isotropic subspaces have dimension $n - 1$.

Structure of orthogonal groups, odd characteristic

Any element of any orthogonal group has determinant ± 1. The subgroup of index 2 consisting of matrices of determinant 1 is the *special orthogonal group*.

The subgroup of scalars has order 2. The resulting *projective special orthogonal group* is *NOT* simple in general. There is (usually) a further subgroup of index 2, which is not so easy to describe.

The spinor norm

In general, orthogonal groups are generated by reflections:

$$r_v : x \mapsto x - 2\frac{B(x, v)}{B(v, v)}v.$$

The reflections have determinant -1, so the special orthogonal group is generated by even products of reflections.

The reflections are of two types: the reflecting vector either has norm a square in F, or a non-square. The subgroup of even products which contain an even number of each type has index 2 (this is NOT obvious!), and is called $\Omega_n(q)$. The projective version $P\Omega_n(q)$ is simple, provided $n \geq 5$.

Orthogonal groups, characteristic 2

These are defined as the isometry groups of non-degenerate *quadratic forms*. This means that the associated bilinear form is non-singular, so the dimension is even. The determinant is always 1. The only scalar in the orthogonal group is 1. Spinor norms have no meaning. But still the orthogonal groups are not simple.

The quasideterminant

If $Q(v) = 1$, the *orthogonal transvection* in v is the map

$$t_v : x \mapsto x + B(x, v)v.$$

In fact, the orthogonal group is generated by these.

There is a subgroup of index 2 consisting of the even products of orthogonal transvections. (This is *NOT* obvious.) This subgroup is simple provided $n \geq 6$.

Small-dimensional orthogonal groups What about dimensions up to 4? In dimension 2, orthogonal groups are dihedral.

- $PSO_3(q) \cong PGL_2(q)$,
- $PSO_4^+(q) \cong (PSL_2(q) \times PSL_2(q)) \cdot 2$,
- $PSO_4^-(q) \cong PSL_2(q^2) \cdot 2$.

Indeed, we can go further:

- $PSO_5(q) \cong PSp_4(q) \cdot 2$, an extension by an automorphism which multiplies the alternating form by a non-square.
- $PSO_6^+(q) \cong PSL_4(q) \cdot 2$, an extension by the "duality" automorphism $M \mapsto (M^T)^{-1}$. This isomorphism is sometimes known as the *Klein correspondence*.
- $PSO_6^-(q) \cong PSU_4(q) \cdot 2$, an extension by the field automorphism $x \mapsto x^q$ (applied to each matrix entry, in the case of the standard inner product).

8. Lie Theory

Lie algebras A *Lie algebra* is a vector space with a product, usually written $[x, y]$, satisfying $[y, x] = -[x, y]$ and

$$[[x, y], z] + [[y, z], x] + [[z, x], y] = 0,$$

the *Jacobi identity*.

The canonical example is the vector space of $n \times n$ matrices of trace 0, with $[x, y] = xy - yx$. This is called $\mathfrak{sl}_n$, and corresponds to the *group SL_n* of matrices of determinant 1. Similarly, there are Lie algebras corresponding to the symplectic and orthogonal groups. Over $\mathbb{C}$ we may take the symplectic algebra to consist of symmetric matrices and the orthogonal algebra to consist of anti-symmetric matrices.

The Jacobi identity implies that the map $d = d_x : y \mapsto [x, y]$ satisfies

$$d([y, z]) = [d(y), z] + [y, d(z)],$$

which is the defining property of a *derivation* of an algebra. Thus a Lie algebra acts on itself as an algebra of derivations.

Conversely, the derivations of any algebra form a Lie algebra under the Lie product $[d, e] = d \circ e - e \circ d$.

Simple Lie algebras

The *simple Lie algebras* over $\mathbb{C}$ are

- A_n, also known as $\mathfrak{sl}_{n+1}$,
- B_n, also known as $\mathfrak{so}_{2n+1}$,
- C_n, also known as $\mathfrak{sp}_{2n}$,
- D_n, also known as $\mathfrak{so}_{2n}$,
- Five exceptional algebras, G_2, F_4, E_6, E_7, and E_8.

Two of the exceptional Lie algebras may be described as algebras of derivations, where a *derivation* is a map d satisfying $d(a \cdot b) = d(a) \cdot b + a \cdot d(b)$. Indeed G_2 is the *algebra of derivations* of the *octonion algebra* (Cayley numbers) which is described in Section 9, and F_4 is the *algebra of derivations* of the *exceptional Jordan algebra* which is described in Section 10.

Coxeter–Dynkin diagrams

Each of the simple Lie algebras over $\mathbb{C}$ is associated with a diagram, which encodes the structure of the algebra. The number of nodes in the diagram is n, and the restrictions on n in the infinite families are simply to avoid repetitions.

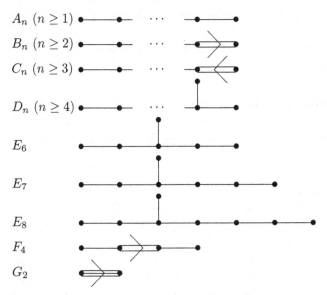

Chevalley groups

There is a general construction of (usually) simple groups, one for each finite field and each diagram. The A_n diagram leads to $PSL_{n+1}(q)$, while the B_n and D_n diagrams lead to $P\Omega_{2n+1}(q)$ and $P\Omega_{2n}^+(q)$. The C_n diagram leads to $PSp_{2n}(q)$.

Corresponding to the exceptional Lie algebras are some exceptional groups of Lie type:

- $G_2(q)$ constructed by L. E. Dickson around 1901, fixing a *cubic form* on a 7-space.
- $E_6(q)$ constructed by L. E. Dickson around 1905, fixing a *cubic form* on a 27-space.
- $F_4(q)$ constructed by C. Chevalley around 1955, acting on a *Lie algebra* of dimension 52.
- $E_7(q)$, acting on a Lie algebra of dimension 133.
- $E_8(q)$, acting on a Lie algebra of dimension 248.

Twisted groups

- The A_n diagram has an automorphism reversing the order of the nodes. This gives rise to the *unitary groups* by a kind of *twisting* operation.
- The D_n diagram has an automorphism swapping the two branches of length 1. This gives rise to the *orthogonal groups of minus type*.
- The E_6 diagram has an automorphism, giving rise to groups called $^2E_6(q)$.
- The D_4 diagram has an automorphism *of order* 3, giving rise to groups called $^3D_4(q)$.

The Suzuki and Ree groups Some of the diagrams have automorphisms only if we ignore the arrows. For reasons we won't go into, this makes sense only if the characteristic of the field is equal to the multiplicity of the edge.

- The *Suzuki groups* $Sz(2^{2n+1}) = {}^2B_2(2^{2n+1})$.
- The *small Ree groups* $R(3^{2n+1}) = {}^2G_2(3^{2n+1})$.
- The *large Ree groups* $R(2^{2n+1}) = {}^2F_4(2^{2n+1})$.

These turn out to be simple if and only if $n \geq 1$. In the case $n = 0$, there is a normal subgroup of index p, where p is the characteristic of the field. The group $Sz(2)$ has order 20. The group $R(3)' \cong PSL_2(8)$ has index 3 in $R(3)$. The group $R(2)'$ has index 2 in $R(2)$ and is a simple group not appearing elsewhere in the classification.

9. Octonions and G_2

Quaternions Hamilton's *quaternions* are defined as

$$\mathbb{H} = \mathbb{R}[i, j, k],$$

where

- $i^2 = j^2 = k^2 = -1$,
- $ij = k = -ji$, $jk = i = -kj$, $ki = j = -ik$.

There is an *involution*

$$^- : a + bi + cj + dk \mapsto a - bi - cj - dk,$$

called *quaternion conjugation*, and a *norm* $N(q) = q\bar{q}$ which satisfies $N(xy) = N(x)N(y)$.

Octonions The octonions, or Cayley numbers, are eight-dimensional over $\mathbb{R}$, and may be defined as $\mathbb{O} = \mathbb{R}[i_0, i_1, \ldots, i_6]$, with subscripts read modulo 7 and a product defined so that i_t, i_{t+1}, i_{t+3} multiply like i, j, k in the quaternions. Then $(i_0 i_1)i_2 = i_3 i_2 = -i_5$ but $i_0(i_1 i_2) = i_0 i_4 = i_5$ so $\mathbb{O}$ is *non-associative*.

It still has an *involution*

$$^- : a + \sum_{t=0}^{6} b_t i_t \mapsto a - \sum_{t=0}^{6} b_t i_t$$

and a *norm* $N(x) = x\bar{x}$ which satisfies $N(xy) = N(x)N(y)$.

$G_2(q)$, for q odd A corresponding octonion algebra exists with coefficients in any finite field of odd characteristic. $G_2(q)$ is the group of linear maps which preserve the *norm* and the *multiplication*. (In fact, it can be shown that if the multiplication is fixed, then so is the norm.) In particular, $G_2(q)$ is inside the orthogonal group $O_8^+(q)$, and fixes 1, so is inside $O_7(q)$.

By counting generating sets for $\mathbb{O}$ equivalent to (i_0, i_1, i_2) we can show that

$$|G_2(q)| = q^6(q^6 - 1)(q^2 - 1).$$

The construction is more complicated in characteristic 2. In fact, $G_2(q)$ is simple for all $q > 2$, while $G_2(2) \cong PSU_3(3).2$, where the automorphism is the field automorphism of $\mathbb{F}_9$.

10.   Exceptional Jordan Algebras and F_4

The Moufang identity   The octonions satisfy the *Moufang identity*

$$(xy)(zx) = (x(yz))x,$$

which substitutes for the associative law in some ways. In particular, if $xyz = 1$ and u satisfies $u\bar{u} = 1$ then

$$
\begin{aligned}
((ux)(yu))(\bar{u}z\bar{u}) &= (u(xy)u)(\bar{u}z\bar{u}) \\
&= u(xy)u\bar{u}z\bar{u} \\
&= u(xy)z\bar{u} = 1.
\end{aligned}
$$

Therefore, the triple of maps

$$(L_u, R_u, B_u) : (x, y, z) \mapsto (ux, yu, \bar{u}z\bar{u})$$

preserves the property that $xyz = 1$.

Triality   Such a triple (α, β, γ) of maps is called an *isotopy*. There are exactly two isotopies for each $\alpha \in \Omega_8^+(q)$, so the isotopies generate a *double cover* of the orthogonal group, called the *spin group*.

    If (α, β, γ) is an isotopy, then (β, γ, α) is an isotopy. This is known as the *triality automorphism* of $P\Omega_8^+(q)$. The centraliser of the triality automorphism is the set of isotopies of the form (α, α, α). This is none other than the automorphism group of the octonions, that is $G_2(q)$.

Jordan algebras   Jordan algebras were introduced by physicists as a way to axiomatise the matrix product $A \circ B = (\frac{1}{2})(AB + BA)$, in an attempt to find a suitable model for quantum mechanics. The essential axiom is the *Jordan identity*

$$((A \circ A) \circ B) \circ A = (A \circ A) \circ (B \circ A).$$

There is one infinite family of simple Jordan algebras corresponding to each of the infinite families of simple Lie algebras.

    It turned out that there was only one new algebra, of dimension 27. This seemed to be of little use for quantum mechanics, but very interesting for group theory.

The exceptional Jordan algebra Take 3×3 Hermitian matrices over the octonions, that is, matrices of the form

$$\begin{pmatrix} a & C & \overline{B} \\ \overline{C} & b & A \\ B & \overline{A} & c \end{pmatrix},$$

where a, b, c are real and A, B, C are octonions. The Jordan product of two such matrices is still Hermitian.

There is a corresponding *exceptional Jordan algebra* with coefficients in any field of characteristic not 2 or 3. The construction is more complicated in characteristics 2 and 3.

$F_4(q)$ in characteristic not 2 or 3 $F_4(q)$ may be defined as the automorphism group of the exceptional Jordan algebra over F_q. To calculate its order, we count the *primitive idempotents*, which are defined as elements e with $e \circ e = e$ and having trace 1.

There are $q^8(q^8 + q^4 + 1)$ of them, and the stabiliser of one of them is a double cover of $SO_9(q)$. Hence,

$$|F_4(q)| = q^{24}(q^{12} - 1)(q^8 - 1)(q^6 - 1)(q^2 - 1).$$

$E_6(q)$ in characteristic not 2 or 3 Surprisingly, the *determinant* of the 3×3 Hermitian octonion matrices makes sense!

$$\det \begin{pmatrix} a & C & \overline{B} \\ \overline{C} & b & A \\ B & \overline{A} & c \end{pmatrix} = abc - a A\overline{A} - b B\overline{B} - c C\overline{C} + (AB)C + \overline{(AB)C},$$

The group of linear maps which preserve this *cubic form* is (modulo scalars) $E_6(q)$.

The order of $E_6(q)$ The notion of *rank* of Hermitian octonion matrices also makes sense, though needs care to define. It can be shown that $E_6(q)$ acts transitively on the matrices of determinant 1.

One of these is the identity matrix, whose stabiliser is $F_4(q)$. Hence, we get the order of $E_6(q)$ (modulo scalars of order $(3, q - 1)$):

$$\frac{q^{36}(q^{12} - 1)(q^9 - 1)(q^8 - 1)(q^6 - 1)(q^5 - 1)(q^2 - 1)}{(3, q - 1)}.$$

11.   Mathieu Groups

The hexacode $\mathbb{F}_4 = \{0, 1, \omega, \overline{\omega}\}$ is the field of order 4. Take six coordinates, grouped into three pairs. Let the *hexacode* $\mathcal{C}$ be the 3-space spanned by

$$(\ \omega \ \overline{\omega} | \overline{\omega} \ \omega | \overline{\omega} \ \omega \ )$$
$$(\ \overline{\omega} \ \omega | \omega \ \overline{\omega} | \overline{\omega} \ \omega \ ).$$
$$(\ \overline{\omega} \ \omega | \overline{\omega} \ \omega | \omega \ \overline{\omega} \ )$$

This is invariant under scalar multiplications (of order 3), permuting the three pairs, and reversing two of the three pairs, which together generate a symmetry group $3 \times S_4$. This group has four orbits on the 63 non-zero vectors in the code:

- 6 of shape $(11|\omega\omega|\overline{\omega}\overline{\omega})$;
- 9 of shape $(11|11|00)$;
- 12 of shape $(\omega\overline{\omega}|\omega\overline{\omega}|\omega\overline{\omega})$;
- 36 of shape $(01|01|\omega\overline{\omega})$.

The full automorphism group of this *code* is $3A_6$, obtained by adjoining the map

$$(ab|cd|ef) \mapsto (\omega a, \overline{\omega} b | cf | de).$$

We can extend to $3S_6$ by mapping

$$(ab|cd|ef) \mapsto (\overline{a}\overline{b}|\overline{c}\overline{d}|\overline{f}\overline{e}).$$

The binary Golay code   Take 24 coordinates (for a vector space over $\mathbb{F}_2$), corresponding to $0, 1, \omega, \overline{\omega}$ in each of the six coordinates of the hexacode:

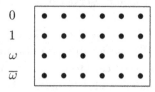

For any given $\mathbb{F}_2^{24}$-vector, and for each column, add up the (entry 0 or 1)×(row label 0, 1, ω or $\overline{\omega}$). Impose the condition that these six sums must form a hexacode word. Also, decree that the parity of each column equals the parity of the top row (even or odd). The set of all vectors which satisfy these conditions is the Golay code.

Some Golay code words

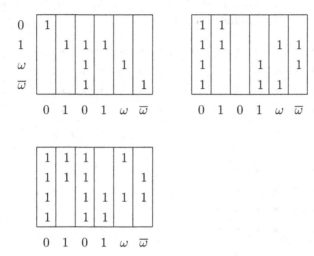

The weight distribution For each of the 64 hexacode words, there are 2^5 even and 2^5 odd *Golay codewords*, making $2^{12} = 4096$ altogether. The 32 odd words are 6 of weight 8, 20 of weight 12, and 6 of weight 16. The 32 even words are:

- For the hexacode word 000000, one word each of weight 0 or 24 and 15 words each of weight 8 or 16.
- For each of the 45 hexacode words of weight 4, 8 Golay code words of each weight 8 or 16, and 16 of weight 12.
- For each of the 18 hexacode words of weight 6, we get all 32 Golay code words of weight 12.

So the full weight distribution is $0^1 8^{759} 12^{2576} 16^{759} 24^1$.

The cosets of the code The Golay code is a vector subspace of dimension 12 in $\mathbb{F}_2^{24}$, so has $2^{12} = 4096$ cosets. The sum (= difference) of two representatives of the same coset is in the code, so has weight at least 8. The identity coset has representative the zero vector (0^{24}). There are 24 cosets with representative of shape $(1, 0^{23})$. There are $24.23/2 = 276$ cosets with representatives of shape $(1^2, 0^{22})$. There are $24.23.22/3.2 = 2024$ cosets with representatives $(1^3, 0^{21})$.

This leaves exactly $1771 = 24.23.22.21/4.3.2.6$ more cosets, so each one has six representatives of shape $(1^4, 0^{20})$. Each of these 1771 cosets defines a *sextet*, that is a partition of the 24 coordinates into 6 sets of 4.

The group $2^6 3S_6$ The 2^6 consists of "adding a hexacode word" to the labels on the rows. The $3S_6$ arises from automorphisms of the hexacode. Here are some examples in diagrammatic form.

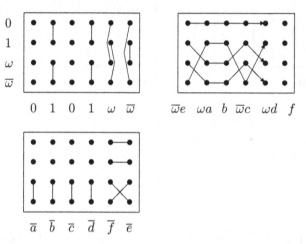

Generating M_{24} The Mathieu group M_{24} may be defined as the automorphism group of the Golay code, that is the set of permutations of 24 points which preserve the set of codewords. Besides the subgroup $2^6 3S_6$, we can show that the permutation

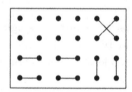

preserves the code.

We can show that M_{24} acts transitively on the 1771 sextets, and that the stabilizer of a sextet is $2^6 3S_6$. Therefore $|M_{24}| = 244823040 = 2^{10} \cdot 3^3 \cdot 5.7.11.23$.

Properties of M_{24} We have seen that every set of 4 points is a part of a sextet. The sextet stabiliser is transitive on the 6 parts (*tetrads*), and a full S_4 of permutations of one part can be achieved. Fixing the four points in the first part is a group $2^4 A_5$ which is transitive on the remaining 20 points. Hence M_{24} acts 5-transitively on the 24 points.

The simplicity of M_{24} can be proved using Iwasawa's Lemma applied to the primitive action on the sextets.

Some subgroups of M_{24}

- The stabiliser of a point has index 24, and is called M_{23}. It has order 10200960.
- The stabiliser of two points is M_{22}, and has order 443520.
- The stabiliser of three points has order 20160, and is in fact $PSL_3(4)$, acting 2-transitively on the $4^2 + 4 + 1 = (4^3 - 1)/(4 - 1)$ points of the *projective plane* of order 4.
- The stabiliser of an *octad* (a codeword of weight 8) has the shape $2^4 A_8$, acting as A_8 on the 8 coordinates, and as $AGL_4(2)$ on the remaining 16.
- The stabiliser of a *dodecad* (a codeword of weight 12) has order $|M_{24}|/2576 = 95040$ and is called M_{12}. It acts 5-transitively on the 12 points in the dodecad.
- The stabiliser of a point in M_{12} is called M_{11} and has order 7920.

The five groups M_{11}, M_{12}, M_{22}, M_{23}, and M_{24} are called the *Mathieu groups*, after Mathieu who discovered them in the 1860s and 1870s.

12. The Leech Lattice and the Conway Groups

The Leech lattice Consider the set of integer vectors $(x_1, \ldots, x_{24}) \in \mathbb{Z}^{24}$ which satisfy

- $x_i \equiv m$ mod 2, that is *either* all the coordinates are even, *or* they are all odd;
- $\sum_{i=1}^{24} x_i \equiv 4m$ mod 8, and
- for each k, the set $\{i \mid x_i \equiv k \bmod 4\}$ is in the Golay code.

This set of vectors, with the inner product $(x_i) \cdot (y_i) = \frac{1}{8} \sum_{i=1}^{24} x_i y_i$, is called the *Leech lattice*.

The group $2^{12} M_{24}$ The lattice is clearly invariant under the group M_{24} acting by permuting the coordinates.

It is also invariant under changing sign on the coordinates corresponding to a Golay code word. These sign changes form an elementary Abelian group of order 2^{12}, that is a direct product of 12 copies of C_2. This is normalised by M_{24}.

Together, these generate a group of shape $2^{12} M_{24}$ of order

$$2^{12}|M_{24}| = 4096.244823040 = 1002795171840.$$

The minimal vectors We classify the vectors of minimal norm in the lattice as follows. If the coordinates are odd, assume that all coordinates are $\equiv 1 \bmod 4$. Since $24 \equiv 0 \bmod 8$, but the sum of the coordinates must be 4 mod 8, the smallest norm is achieved when the vector has shape $(-3, 1^{23})$. The norm is then $\frac{1}{8}(9 + 23) = 4$.

If the coordinates are even, and some are 2 mod 4, then at least 8 of them are 2 mod 4, and the minimal norm is achieved with vectors of shape $(2^8, 0^{16})$. Otherwise they are all divisible by 4, and the minimal norm is achieved with vectors of shape $(4^2, 0^{22})$.

Under the action of the group $2^{12} M_{24}$, there are $24 \cdot 2^{12} = 98304$ images of the vector $(-3, 1^{23})$. For each of the 759 octads, there are 2^7 vectors of shape $(\pm 2^8, 0^{16})$, making $2^7 \cdot 759 = 97152$ in all. There are $(24.23/2) \cdot 2^2 = 1104$ vectors of shape $(\pm 4^2, 0^{22})$. So in total there are $98304 + 97152 + 1104 = 196560$ vectors of norm 4 in the Leech lattice.

Vectors of norms 6 and 8 Similar arguments can be used to classify the vectors of norms 6 and 8. The 16773120 vectors of norm 6 are

- $2^{11} \cdot 2576$ of shape $(\pm 2^{12}, 0^{12})$,
- 24.2^{12} of shape $(5, 1^{23})$,
- $(24.23.22/3.2).2^{12}$ of shape $(-3^3, 1^{21})$,
- $759.16.2^8$ of shape $(2^7, -2, 4)$.

There are 398034000 vectors of norm 8.

Congruence classes modulo 2 Define two vectors in the Leech lattice to be *congruent mod 2* if their difference is twice a Leech lattice vector. Clearly, every vector is congruent mod 2 to its negative.

If x is congruent to y, we may assume $x \cdot y$ is positive so that

$$16 \leq (x - y) \cdot (x - y) \leq x \cdot x + y.y.$$

Therefore, the only non-trivial congruences between vectors of norm ≤ 8 are between perpendicular vectors of norm 8.

Thus, the vectors of norm at most 8 account for at least

$$\frac{1 + 196560}{2} + \frac{16773120}{2} + \frac{398034000}{48},$$

congruence classes. But this $= 2^{24}$, so these are all.

Crosses In particular, each congruence class of norm 8 vectors consists of 24 pairs of mutually orthogonal vectors, called a *cross*. Thus, there are 8292375 crosses.

By verifying that the linear map which acts on each column as

$$\frac{1}{2} \begin{pmatrix} 1 & -1 & -1 & -1 \\ -1 & -1 & 1 & 1 \\ -1 & 1 & -1 & 1 \\ -1 & 1 & 1 & -1 \end{pmatrix},$$

preserves the lattice, we can show that the automorphism group of the lattice acts transitively on the crosses. Therefore, the stabiliser of any cross (or coordinate frame) is $2^{12}M_{24}$.

Conway's group Hence, the automorphism group of the Leech lattice has order

$$8292375 \cdot |2^{12}M_{24}| = 8315553613086720000.$$

It has a centre of order 2, consisting of the scalars ± 1. Factoring out the centre gives a group Co_1, namely, *Conway's first group* of order

$$4157776806543360000.$$

Using the primitive action on the 8292375 crosses, and Iwasawa's Lemma, we can easily prove that Co_1 is simple.

Some subgroups of the Conway group

- The automorphism group of the Leech lattice is transitive on the norm 4 vectors, and the stabiliser of a norm 4 vector is Co_2, *Conway's second group*.
- Similarly, the stabiliser of a norm 6 vector is Co_3, *Conway's third group*.
- The stabiliser of two norm 4 vectors whose sum has norm 6 is the *McLaughlin group McL*.
- The stabiliser of two norm 6 vectors whose sum has norm 4 is the *Higman–Sims group HS*.
- The centraliser of a fixed-point-free element of order 3 is a sixfold cover of the *Suzuki group Suz*.
- The centraliser of a fixed-point-free element of order 5 contains a double cover of the *Hall–Janko group J_2*.

13. Further Reading

A much fuller treatment of all the finite simple groups, along the same
general lines as in this chapter, can be found in my book [7]. There are
many books which cover some particular types of simple groups, at various
levels. For example, there are Sagan's book on the symmetric groups [5],
which contains a wealth of information, and Carter's classic book [2] on the
groups of Lie type, both classical and exceptional. For a more introductory-
level text on the classical groups, see for example Grove [4]. For exceptional
groups, see Springer and Veldkamp [6], and for sporadic groups, Griess [3].
The Suzuki and Ree groups are constructed without Lie theory in my paper
[8]. Finally, for detailed examples, one should always have at hand a copy
of the ATLAS [1].

References

[1] J. H. Conway, R. T. Curtis, S. P. Norton, R. A. Parker and R. A. Wilson, *An
     Atlas of Finite Groups*, Oxford University Press, Oxford, 1985.
[2] R. W. Carter, *Simple groups of Lie type*, Wiley, 1972.
[3] R. L. Griess, *Twelve sporadic groups*, Springer MM, 1998.
[4] L. C. Grove, *Classical groups and geometric algebra*, Amer. Math. Soc., 2002.
[5] B. E. Sagan, *The symmetric group*, Springer, 2001.
[6] T. A. Springer and F. D. Veldkamp, *Octonions, Jordan algebras and excep-
     tional groups*, Springer MM, 2000.
[7] R. A. Wilson, *The Finite Simple Groups*, Springer GTM 251, 2009.
[8] R. A. Wilson, On the simple goups of Suzuki and Ree, *Proc. London Math.
     Soc.* **107** (2013), 680–712.

Chapter 3

Introduction to Representations of Algebras and Quivers

Anton Cox

Department of Mathematics,
City University London,
Northampton Square,
London EC1V 0HB, UK
A.G.Cox@city.ac.uk

This chapter introduces the basic representation theory of finite-dimensional algebras. We outline the basic structure theorems for finite-dimensional algebras such as Artin–Wedderburn and Krull–Schmidt. We then introduce projective and injective modules, and show how the study of representations of finite-dimensional algebras can be reduced to the study of bound quiver algebras.

1. Algebras and Modules

We will be interested in the representation theory of finite-dimensional algebras defined over a field. We begin by recalling certain basic definitions concerning fields.

Definition 1. A field k is *algebraically closed* if every non-constant polynomial with coefficients in k has a root in k. A field has *characteristic p* if p is the smallest positive integer such that

$$\sum_{i=1}^{p} 1 = 0.$$

If there is no such p then the field is said to have *characteristic 0*. Henceforth, k will denote some field.

1.1. *Associative algebras*

Definition 2. An *algebra over* k, or *k-algebra* is a k-vector space A with a bilinear map

$$A \times A \longrightarrow A,$$
$$(x, y) \longmapsto xy.$$

We say that the algebra is *associative* if for all $x, y, z \in A$ we have

$$x(yz) = (xy)z.$$

An algebra A is *unital* if there exists an element $1 \in A$ such that $1x = x1 = x$ for all $x \in A$. Such an element is called the *identity* in A. (Note that such an element is necessarily unique.) We say that an algebra is *finite-dimensional* if the underlying vector space is finite-dimensional. An algebra A is *commutative* if $xy = yx$ for all $x, y \in A$.

It is common to abuse terminology and take algebra to mean an associative unital algebra, and we will follow this convention. There are several important classes of non-associative algebras (for example Lie algebras) but we shall not consider them here. *Thus, all algebras we consider will be associative and unital.*

Example 1. (a) Let $k[x_1, \ldots, x_n]$ denote the vector space of polynomials in the (commuting) variables $x_1, \ldots, x_n$. This is an infinite dimensional commutative algebra with multiplication given by the usual multiplication of polynomials, and identity given by the trivial polynomial 1.

(b) Let $k\langle x_1, \ldots, x_n \rangle$ denote the vector space of polynomials in the *non-commuting* variables $x_1, \ldots, x_n$. A general element is of the form $\sum_{i=1}^{n} \lambda_i w_i$ for some n where for each i, $\lambda_i \in k$ and $w_i = x_{i_1}^{a_1} x_{i_2}^{a_2} \ldots x_{i_t}^{a_t}$ for some t. Given two elements $\sum_{i=1}^{n} \lambda_i w_i$ and $\sum_{i=1}^{m} \lambda_i' w_i'$ the product is defined to be the element

$$\sum_{i=1}^{n} \sum_{j=1}^{m} \lambda_i \lambda_i' w_i w_j',$$

where $w_i w_j$ denotes the element obtained from w_i and w_j by concatenation. This is an infinite dimensional associative algebra with identity given by the trivial polynomial 1. If $n > 1$ then the algebra is non-commutative.

(c) Given a group G, we denote by kG the *group algebra* obtained by considering the vector space of formal linear combinations of group elements. Given two elements $\sum_{i=1}^{n} \lambda_i g_i$ and $\sum_{i=1}^{m} \mu_i h_i$ with $\lambda_i, \mu_i \in k$ and $g_i, h_i \in G$ we define the product to be the element

$$\sum_{i=1}^{n} \sum_{j=1}^{m} \lambda_i \mu_j g_i h_j.$$

The identity element is the identity element $e \in G$ regarded as an element of kG. The algebra kG is finite-dimensional if and only if G is a finite group, and is commutative if and only if G is Abelian.

(d) The set $M_n(k)$ of $n \times n$ matrices with entries in k is a finite-dimensional algebra, the *matrix algebra*, with the usual matrix multiplication, and identity element the matrix I. If $n > 1$, it is non-commutative. Equivalently, let V be an n-dimensional k-vector space, and consider the *endomorphism algebra*

$$\mathrm{End}_k(V) = \{f : V \longrightarrow V \mid f \text{ is } k\text{-linear}\}.$$

This is an algebra with multiplication given by composition of functions. Fixing a basis for V the elements of $\mathrm{End}_k(V)$ can be written in terms of matrices with respect to this basis, and in this way we can identify $\mathrm{End}_k(V)$ with $M_n(k)$.

(e) If A is an algebra then so is A^{op}, the *opposite algebra*, which equals A as a vector space, but with multiplication map $(x, y) \longmapsto yx$.

As usual in algebra, we are not just interested in objects (in this case algebras), but also in functions between them which respect the underlying structures.

Definition 3. A *homomorphism* between k-algebras A and B is a linear map $\phi : A \longrightarrow B$ such that $\phi(1) = 1$ and $\phi(xy) = \phi(x)\phi(y)$ for all $x, y \in A$. This is an *isomorphism* precisely when the linear map is a bijection.

Example 2. If G is a group then $kG \cong (kG)^{op}$.

Definition 4. Given an algebra A, a *subalgebra* of A is a subspace S of A containing 1, such that for all $x, y \in S$ we have $xy \in S$. A *left (respectively right) ideal* in A is a subspace I of A such that for all $x \in I$ and $a \in A$ we have $ax \in I$ (respectively $xa \in I$). If I is a left and a right ideal then we say that I is an *ideal* in A.

Example 3. (a) If H is a subgroup of a group G, then kH is a subalgebra of kG.

(b) Given two algebras A and B, and a homomorphism $\phi : A \longrightarrow B$, the set $\mathrm{im}(\phi)$ is a subalgebra of B, while $\ker(\phi)$ is an ideal in A.

Idempotents play a crucial role in the analysis of algebras.

Definition 5. An element $e \in A$ is an *idempotent* if $e^2 = e$. Two idempotents e_1 and e_2 in A are *orthogonal* if

$$e_1 e_2 = e_2 e_1 = 0.$$

An idempotent e is called *primitive* if it cannot be written in the form $e = e_1 + e_2$ where e_1 and e_2 are non-zero orthogonal idempotents. An idempotent e is *central* if $ea = ae$ for all $a \in A$.

1.2.   *Modules*

Representation theory is concerned with the study of the way in which certain algebraic objects (in our case, algebras) act on vector spaces. There are two ways to express this concept; in terms of representations or (in more modern language) in terms of modules.

Definition 6. Given an algebra A over k, a *representation* of A is an algebra homomorphism

$$\phi : A \longrightarrow \mathrm{End}_k(M)$$

for some vector space M. A *left A-module* is a k-vector space M together with a bilinear map $A \times M \longrightarrow M$, which we will denote by $(a, m) \longmapsto am$, such that for all $m \in M$ and $x, y \in A$ we have $1m = m$ and $(xy)m = x(ym)$. Similarly, a *right A-module* is a k-vector space M and a bilinear map $\phi : M \times A \longrightarrow M$ such that $m1 = m$ and $m(xy) = (mx)y$ for all $m \in M$ and $x, y \in A$. We will adopt the convention that all modules are left modules unless stated otherwise.

Definition 7. An A-module is *finite-dimensional* if it is finite-dimensional as a vector space. An A-module M is *generated* by a set $\{m_1 : i \in I\}$ (where I is some index set) if every element m of M can be written in the form

$$m = \sum_{i \in I} a_i m_i$$

for some $a_i \in A$. We say that M is *finitely generated* if it is generated by a finite set of elements. If A is a finite-dimensional algebra then M is finitely generated if and only if M is finite-dimensional.

Lemma 1. (a) *There is a natural equivalence between left (respectively right) A-modules and right (respectively left) A^{op}-modules.*

(b) *There is a natural equivalence between representations of A and left A-modules.*

Proof. We give the correspondence in each case; details are left to the reader. Given a left module M for A with bilinear map $\phi : A \times M \longrightarrow M$, define a right A^{op}-module structure on M via the map $\phi' : M \times A \longrightarrow M$ given by $\phi'(m,x) = \phi(x,m)$. It is easy to verify that ϕ is an A^{op}-homomorphism.

Given a representation $\phi : A \longrightarrow \text{End}_k(M)$ of A we define an A-module structure on M by setting $am = \phi(a)(m)$ for all $a \in A$ and $m \in M$. Conversely, given an A-module M, the map $M \longrightarrow M$ given by $m \longmapsto rm$ is linear, and gives the desired representation $\phi : A \longrightarrow \text{End}_k(M)$. □

Definition 8. A *homomorphism* between A-modules M and N is a linear map $\phi : M \longrightarrow N$ such that $\phi(am) = a\phi(m)$ for all $a \in A$ and $m \in M$. This is an *isomorphism* precisely when the linear map is a bijection.

Definition 9. Given an A-module M, a *submodule* of M is a subspace N of M such that for all $n \in N$ and $a \in A$ we have $an \in N$. (Note that N is an A-module in its own right.) The quotient space

$$M/N = \{m + N : m \in M\},$$

(under the relation $m + N = m' + N$ if and only if $m - m' \in N$) has an A-module structure given by $a(m+N) = am+N$, and is called the *quotient* of M by N.

Example 4. (a) The algebra A is a (left or right) A-module, with respect to the usual multiplication map on A. If I is a left ideal of A then I is a submodule of the left module A. Further, A/I can be given an algebra structure such that the natural map from A onto A/I is an algebra surjection, and any A/I-module M has a natural A-module structure.

(b) If $A = k$ then A-modules are just k-vector spaces.

(c) If $A = k[x_1, \ldots, x_n]$ then an A-module is a k-vector space M together with commuting linear transformations $\alpha_i : M \longrightarrow M$ (where α_i describes the action of x_i).

(d) Every A-module M has M and the empty vector space 0 as submodules.

Lemma 2 (Isomorphism Theorem). *If M and N are A-modules and $\phi : M \longrightarrow N$ is a homomorphism of A-modules then*

$$\text{im}(\phi) \cong M/\ker(\phi),$$

as A-modules.

Proof. Copy the proof for linear maps between vector spaces, noting that the additional structure of a module is preserved. □

Definition 10. If an A-module M has submodules L and N such that $M = L \oplus N$ as a vector space then we say that M is the *direct sum* of L and N. A module M is *indecomposable* if it is not the direct sum of two non-zero submodules (and is *decomposable* otherwise). A module M is *simple* (or *irreducible*) if M has no submodules except M and 0.

For vector spaces, the notions of indecomposability and irreducibility coincide. However, this is not the case for modules in general.

Example 5. Let C_2 denote the cyclic group with elements $\{1, g\}$, and consider the two-dimensional kC_2-module M with basis $\{m_1, m_2\}$ where $gm_1 = m_2$ and $gm_2 = m_1$. If $M = N_1 \oplus N_2$ with N_1 and N_2 non-zero then each N_i is the span of a vector of the form $\lambda_1 m_1 + \lambda_2 m_2$ for some $\lambda_1, \lambda_2 \in k$. Applying g we deduce that $\lambda_1 = \pm\lambda_2$, and hence N_i must be the span of $m_1 - m_2$ or $m_1 + m_2$. But $N_1 = N_2$ if k has characteristic 2, which contradicts our assumption. Thus, M is never irreducible, but is indecomposable if and only if the characteristic of k is 2.

The representation theory of A and of A^{op} are closely related.

Definition 11. Let M be a finite-dimensional (left) A-module. Then the *dual module* M^* is the dual vector space $\mathrm{Hom}_k(M, k)$ with a right A-module action given by $(\phi a)(m) = \phi(am)$ for all $a \in A$, $m \in M$ and $\phi \in \mathrm{Hom}_k(M, k)$. By Lemma 1 this gives M^* the structure of a left A^{op}-module.

Taking the dual of an A^{op}-module gives an A-module, and it is easy to verify (as for vector spaces) that:

Lemma 3. *For any finite-dimensional A-module M we have $M^{**} \cong M$.*

1.3.   *Quivers*

Definition 12. A *quiver* Q is a directed graph. We will denote the set of vertices by Q_0, and the set of edges (which we call *arrows*) by Q_1. If Q_0 and Q_1 are both finite then Q is a *finite* quiver. The *underlying graph* $\bar{Q}$ of a quiver Q is the graph obtained from Q by forgetting all orientations of edges.

A *path of length n* in Q is a sequence $p = \alpha_1 \alpha_2 \ldots \alpha_n$ where each α_i is an arrow and α_i starts at the vertex where α_{i+1} ends. For each vertex i, there is a path of length 0, which we denote by ϵ_i. A quiver is *acyclic* if the only paths which start and end at the same vertex have length 0, and *connected* if $\bar{Q}$ is a connected graph.

Example 6. (a) For the quiver Q given by

the set of paths of length greater than 1 is given by

$$\{\beta^{n+2}, \beta^{n+1}\alpha, \gamma\beta^{n+1}, \delta\beta^{n+1}, \gamma\beta^n\alpha, \delta\beta^n\alpha : n \geq 0\}.$$

(b) For the quiver Q given by

the set of paths corresponds to words in α and β (along with the trivial word).

(c) For the quiver Q given by

the set of paths is

$$\{\epsilon_1, \epsilon_2, \epsilon_3, \epsilon_4, \alpha, \beta, \gamma, \beta\alpha\}.$$

We would like to associate an algebra to a quiver; however, we need to take a little care.

Definition 13. The *path algebra* kQ of a quiver Q is the k-vector space with basis the set of paths in Q. Multiplication is via concatenation of paths: if $p = \alpha_1\alpha_2\dots\alpha_n$ and $q = \beta_1\beta_2\dots\beta_m$ then

$$pq = \alpha_1\alpha_2\dots\alpha_n\beta_1\beta_2\dots\beta_m,$$

if α_n starts at the vertex where β_1 ends, and is 0 otherwise.

We have not yet checked that the above definition does in fact define an algebra structure on kQ.

Lemma 4. *Let Q be a quiver. Then kQ is an associative algebra. Further, kQ has an identity element if and only if Q_0 is finite, and is finite-dimensional if and only if Q is finite and acyclic.*

Proof. The associativity of multiplication in kQ is straightforward. Next, note that the elements ϵ_i satisfy $\epsilon_i \epsilon_j = \delta_{ij} \epsilon_i$ and hence form a set of orthogonal idempotents. Further, for any path $p \in kQ$ we have $\epsilon_i p = p$ if p ends at vertex i and 0 otherwise (and similarly for $p\epsilon_i$). Hence, if Q_0 is finite then

$$\sum_{i \in Q_0} \epsilon_i p = p \qquad \text{and} \qquad \sum_{i \in Q_0} p\epsilon_i = p$$

and hence,

$$1 = \sum_{i \in Q_0} \epsilon_i$$

is the unit in kQ.

Conversely, suppose that Q_0 is infinite and $1 \in kQ$. Then $1 = \sum \lambda_i p_i$ for some (finite) set of paths p_i and scalars λ_i. Pick a vertex j such that for all i the path p_i does not end at j. Then $\epsilon_j 1 = 0$, which gives a contradiction.

Finally, if Q_0 or Q_1 is not finite then kQ is clearly not finite-dimensional. Given a finite set of vertices with finitely many edges, there are only finitely many paths between them unless the quiver contains a cycle. □

Example 7. Each of the quivers in Example 6 is finite, and so the corresponding kQ contains a unit. However, the path algebras corresponding to 6(a) and 6(b) are not finite-dimensional. Indeed, it is easy to see that the path algebra for (b) is isomorphic to $k\langle x, y \rangle$, under the map taking α to x and β to y. The path algebra for 6(c) is an eight-dimensional algebra.

Because of Lemma 4, we will only consider finite quivers Q, so that the corresponding path algebras are unital.

Definition 14. Given a finite quiver Q, the ideal R_Q of kQ generated by the arrows in Q is called the *arrow ideal* of kQ. Then R_Q^m is the ideal generated by all paths of length m in Q. An ideal I in kQ is called *admissible* if there exists $m \geq 2$ such that

$$R_Q^m \subseteq I \subseteq R_Q^2.$$

If I is admissible then (Q, I) is called a *bound quiver*, and kQ/I is a *bound quiver algebra*.

Note that if Q is finite and acyclic then any ideal contained in R_Q^2 is admissible, as $R_Q^m = 0$ if m is greater than the maximal path length in Q.

Example 8. Let Q be as in Example 6(b), and let $I = \langle \beta\alpha, \beta^2 \rangle$. This is not an admissible ideal in kQ as it does not contain α^m for any $m \geq 1$, and so does not contain R_Q^m for any $m \geq 2$.

Proposition 1. *Let Q be a finite quiver with admissible ideal I in kQ. Then kQ/I is finite-dimensional.*

Proof. As I is admissible there exists $m \geq 2$ such that $R_Q^m \subseteq I$. Hence, there is a surjective algebra homomorphism from kQ/R_Q^m onto kQ/I. But the former algebra is clearly finite-dimensional as there are only finitely many paths of length less than m. □

Definition 15. A *relation* in kQ is a finite linear combination of paths of length at least two in Q such that all paths have the same start vertex and the same end vertex. If $\{\rho_j : j \in J\}$ is a set of relations in kQ such that the ideal generated by the set is admissible then we say that kQ is *bound by the relations*.

Example 9. Consider the quiver in Example 6(a) and the relations

$$\{\gamma\beta^2\alpha - \delta\alpha, \gamma\beta + \delta\beta, \beta^5\}.$$

Any path of length at least 7 must contain β^5, and so Q is bound by this set of relations.

In fact, the above example generalises: it can be shown that any ideal I in R_Q^2 is admissible if it contains each cycle in Q to some power. Further, we have:

Proposition 2. *Let Q be a finite quiver. Every admissible ideal in kQ is generated by a finite sequence of relations in kQ.*

Proof. (Sketch) It is easy to check that every admissible ideal I is finitely generated by some set $\{a_1, \ldots, a_n\}$ (as R_Q^m and I/R_Q^m are finitely generated). However, in general a set of generators for I will not be a set of relations, as the paths in each a_i may not all have the same start vertex and end vertex. However, the non-zero elements in the set

$$\{\epsilon_x a_i \epsilon_y : 1 \leq i \leq n, x, y \in Q_0\},$$

are all relations, and this set generates I. □

1.4. Representations of quivers

Definition 16. Let Q be a finite quiver. A *representation* M of Q over k is a collection of k-vector spaces $\{M_a : a \in Q_0\}$ together with a linear map

$\phi_\alpha : M_a \longrightarrow M_b$ for each arrow $\alpha : a \longrightarrow b$ in Q_1. The representation M is *finite-dimensional* if all the M_a are finite-dimensional.

Definition 17. Given two representations M and M' of a finite quiver Q, a *homomorphism from M to N* is a collection of linear maps $f_i : M_i \longrightarrow M'_i$ such that for each arrow $\alpha : i \longrightarrow j$ we have $\phi'_\alpha f_i = f_j \phi_\alpha$.

When giving examples of representations of quivers, we will usually fix bases of each of the vector spaces, and represent the maps between them by matrices with respect to column vectors in these bases.

Example 10. Consider the quiver

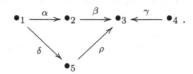

This has a representation

$$k \xrightarrow{\binom{1}{0}} k^2 \xrightarrow{\left(\begin{smallmatrix} 1 & 1 \\ 1 & 0 \end{smallmatrix}\right)} k^2 \xleftarrow{\left(\begin{smallmatrix} 1 & 2 & 1 \\ 0 & 1 & 0 \end{smallmatrix}\right)} k^3 \ .$$

with maps $\binom{0}{1}$ and $\left(\begin{smallmatrix} 0 & 1 \\ 1 & 1 \end{smallmatrix}\right)$ to k^2.

Notice how easy it was to give a representation: there are no compatibility relations to be checked (apart from that the linear maps go between the appropriate dimension) so examples can be easily generated for any path algebra. This is very different from writing down explicit modules for an algebra (in general).

Definition 16 looks rather different from that for an algebra. However, the next lemma shows that representations of Q correspond to kQ-modules in a natural way.

Lemma 5. *Let M be a representation of a finite acyclic quiver Q. Consider the vector space*

$$M' = \bigoplus_{a \in Q_0} M_a.$$

This can be given the structure of a kQ-module by defining for each $\alpha : i \longrightarrow j$ a map $\phi'_\alpha : M \longrightarrow M$ by

$$\phi'_\alpha(m_1, \ldots, m_n) = (0, \ldots, 0, \phi_\alpha(m_i), 0, \ldots 0),$$

where the non-zero entry is in position j, and for each $i \in Q_0$ a map $\epsilon_i : M \longrightarrow M$ by

$$\epsilon_i(m_1, \ldots, m_n) = (0, \ldots, 0, m_i, 0, \ldots, 0),$$

where the non-zero entry is in position i. Conversely, suppose that N is a kQ-module. Then we obtain a representation of Q by setting $N_a = \epsilon_a N$ and defining ϕ_α for $\alpha : a \longrightarrow b$ to be the restriction of the action of $\alpha \in kQ$ to N_a.

Proof. Checking that the above definitions give a kQ-module and a representation of Q respectively is routine. □

We also need the notion of a representation of a bound quiver. Note that we do not need to assume that Q is acyclic here, as admissible ideals guarantee that the associated quotient algebra is finite-dimensional.

Definition 18. Given a path $p = \alpha_1 \alpha_2 \ldots \alpha_n$ in a finite quiver Q from a to b and a representation M of Q we define the linear map ϕ_p from M_a to M_b by

$$\phi_p = \phi_{\alpha_n} \phi_{\alpha_{n-1}} \cdots \phi_{\alpha_1}.$$

If ρ is a linear combination of paths p_i with the same start vertex and the same end vertex then ϕ_ρ is defined to be the corresponding linear combination of the ϕ_{p_i}. Given an admissible ideal I in kQ we say that M is *bound by I* if $\phi_\rho = 0$ for all relations $\rho \in I$.

Example 11. Consider the representation in Example 10. Let $p = \beta\alpha$ and $q = \rho\delta$. Then

$$\phi_p = \begin{pmatrix} 1 & 1 \\ 1 & 0 \end{pmatrix} \begin{pmatrix} 1 \\ 0 \end{pmatrix} = \begin{pmatrix} 1 \\ 1 \end{pmatrix} \qquad \phi_q = \begin{pmatrix} 0 & 1 \\ 1 & 1 \end{pmatrix} \begin{pmatrix} 0 \\ 1 \end{pmatrix} = \begin{pmatrix} 1 \\ 1 \end{pmatrix}$$

and so this representation is bound by the ideal $\langle \beta\alpha - \rho\delta \rangle$.

It is easy to verify that the correspondence between representation of finite acyclic Q and kQ-modules given in Lemma 5 extends to a correspondence between representations of finite Q bound by I and kQ/I-modules.

The language of categories and functors is a very powerful one, and many results in representation theory are best stated in this way. Roughly, a category is a collection of *objects* (e.g., kQ-modules) and *morphisms* (e.g., kQ-homomorphisms), and the idea is to study the category as a whole rather than just the objects or morphisms separately. A *functor* is then a map from one category to another which transports both objects and morphisms in a suitably compatible way. In this language, the above result relating bound representations of Q and kQ/I-modules gives an equivalence between the corresponding categories.

2.   Semisimplicity and Some Basic Structure Theorems

In this section, we will review some of the classical structure theorems for finite-dimensional algebras. In most cases, results will be stated with only a sketch of the proof. Henceforth, we will restrict our attention to finite-dimensional modules.

2.1.   *Simple modules and semisimplicity*

Recall that a simple module is a module S such that the only submodules are S and 0. These form the building blocks out of which all other modules are made:

Lemma 6. *If M is a finite-dimensional A-module then there exists a sequence of submodules*

$$0 = M_0 \subset M_1 \subset \cdots \subset M_n = M,$$

such that M_i/M_{i-1} is simple for each $1 \leq i \leq n$. Such a series is called a composition series *for M.*

Proof. Proceed by induction on the dimension of M. If M is not simple, pick a submodule M_1 of minimal dimension, which is necessarily simple. Now $\dim(M/M_1) < \dim M$, and so the result follows by induction.   □

Theorem 1 (Jordan–Hölder). *Suppose that M has two composition series*

$$0 = M_0 \subset M_1 \subset \cdots \subset M_m = M, \qquad 0 = N_0 \subset N_1 \subset \cdots \subset N_n = M.$$

Then $n = m$ and there exists a permutation σ of $\{1, \ldots n\}$ such that

$$\frac{M_i}{M_{i+1}} \cong \frac{N_{\sigma(i)}}{N_{\sigma(i)+1}}.$$

Proof. The proof is similar to that for groups.   □

Life would be (relatively) straightforward if every module was a direct sum of simple modules.

Definition 19. A module M is *semisimple* (or *completely reducible*) if it can be written as a direct sum of simple modules. An algebra A is *semisimple* if every finite-dimensional A-module is semisimple.

Lemma 7. *If M is a finite-dimensional A-module then the following are equivalent:*
(a) *If N is a submodule of M then there exists L a submodule of M such that $M = L \oplus N$.*
(b) *M is semisimple.*
(c) *M is a (not necessarily direct) sum of simple submodules.*

Proof. (Sketch) Note that (a) implies (b) and (b) implies (c) are clear. For (c) implies (a) consider the set of submodules of A whose intersection with N is 0. Pick one such, L say, of maximal dimension; if $N \oplus L \neq M$ then there is some simple S in M not in $N \oplus L$. But this would imply that $S + L$ has intersection 0 with A, contradicting the maximality of L.   □

Lemma 8. *If M is a semisimple A-module then so is every submodule and quotient module of M.*

Proof. (Sketch) If N is a submodule then $M = N \oplus L$ for some L by the preceding lemma. But then $M/L \cong N$, and so it is enough to prove the result for quotient modules.

If M/L is a quotient module consider the projection homomorphism π from M to M/L. Write M as a sum of simple modules S_i and verify that $\pi(S)$ is either simple or 0. This proves that M/L is a sum of simple modules, and so the result follows from the preceding lemma.   □

To show that an algebra is semisimple, we do not want to have to check the condition for every possible module. Fortunately we have

Proposition 3. *Every finite-dimensional A-module is isomorphic to a quotient of A^n for some n. Hence an algebra A is semisimple if and only if A is semisimple as an A-module.*

Proof. (Sketch) Suppose that M is a finite-dimensional A-module, spanned by some elements $m_1, \ldots, m_n$. We define a map

$$\phi : \oplus_{i=1}^n A \longrightarrow M,$$

by

$$\phi((a_1, \ldots, a_n)) = \sum_{i=1}^{n} a_i m_i.$$

It is easy to check that this is a homomorphism of A-modules, and so by the isomorphism theorem we have that

$$M \cong \frac{\oplus_{i=1}^{n} A}{\ker \phi}.$$

The result now follows from the preceding lemma. $\qquad \square$

For finite groups, we can say exactly when kG is semisimple:

Theorem 2 (Maschke). *Let G be a finite group. Then the group algebra kG is semisimple if and only if the characteristic of k does not divide $|G|$, the order of the group.*

Proof. (Sketch) First suppose that the characteristic of k does not divide $|G|$. We must show that every kG-submodule M of kG has a complement as a module. Clearly, as vector spaces we can find N such that $M \oplus N = kG$. Let $\pi : kG \longrightarrow M$ be the projection map $\pi(m + n) = m$ for all $m \in M$ and $n \in N$. We want to modify π so that it is a module homomorphism, and then show that the kernel is the desired complement.

Define a map $T_\pi : kG \longrightarrow M$ by

$$T_\pi(m) = \frac{1}{|G|} \sum_{g \in G} g(\pi(g^{-1}m)).$$

Note that this is possible as $|G|^{-1}$ exists in k. It is then routine to check that T_π is a kG-module map, and that $K = \ker(T_\pi)$ satisfies $kG = M \oplus K$.

For the reverse implication, consider $w = \sum_{g \in G} g \in kG$. It is easy to check that every element of g fixes w, and hence w spans a one-dimensional submodule M of kG. Now suppose that there is a complementary submodule N of kG, and decompose $1 = e + f$ where e and f are the idempotents corresponding to M and N respectively. We have $e = \lambda w$ for some $\lambda \in k$, and $e^2 = e = \lambda^2 w^2$. It is easy to check that $w^2 = |G|w$ and hence $\lambda w = \lambda^2 |G| w$ which implies that $1 = \lambda |G|$. But this contradicts the fact that $|G| = 0$ in k. $\qquad \square$

The next result will be important in the following section.

Lemma 9. *The algebra $M_n(k)$ is semisimple.*

Proof. Let E_{ij} denote the matrix in $A = M_n(k)$ consisting of zeros everywhere except for the (i, j)th entry, which is 1. We first note that

$$1 = E_{11} + E_{22} + \cdots + E_{nn}$$

is an orthogonal idempotent decomposition of 1, and hence A decomposes as a direct sum of modules of the form AE_{ii}. We will show that these summands are simple.

First, observe that AE_{ii} is just the set of matrices which are zero except possibly in column i. Pick $x \in AE_{ii}$ non-zero; we must show that $Ax = AE_{ii}$. As x is non-zero there is some entry x_{mi} in the matrix x which is non-zero. But then

$$E_{jm}x = x_{mi}E_{ji} \in Ax$$

and hence $E_{ji} \in Ax$ for all $1 \leq j \leq n$. But this implies that $Ax = AE_{ii}$ as required. $\qquad\square$

2.2. *Schur's lemma and the Artin–Wedderburn theorem*

We begin with Schur's lemma, which tells us about automorphisms of simple modules.

Lemma 10 (Schur). *Let S be a simple A-module and $\phi : S \longrightarrow S$ a non-zero homomorphism. Then ϕ is invertible.*

Proof. Let $M = \ker \phi$ and $N = \operatorname{im} \phi$; these are both submodules of S. But S is simple and $\phi \neq 0$, so $M = 0$ and ϕ is injective. Similarly we see that $N = S$, so ϕ is surjective, and hence ϕ is invertible. $\qquad\square$

Lemma 11. *If k is algebraically closed and S is a finite-dimensional simple module with non-zero endomorphism ϕ, then $\phi = \lambda.\operatorname{id}_S$, for some non-zero $\lambda \in k$.*

Proof. As k is algebraically closed and $\dim S < \infty$ the map ϕ has an eigenvalue $\lambda \in k$. Then $\phi - \lambda \operatorname{id}_S$ is an endomorphism of S with non-zero kernel (containing all eigenvectors with eigenvalue λ). Arguing as in the preceding lemma we deduce that $\ker(\phi - \lambda \operatorname{id}_S) = S$, and hence $\phi = \lambda \operatorname{id}_S$. $\qquad\square$

Given an A-module M, we set

$$\operatorname{End}_A(M) = \{\phi : M \longrightarrow M \mid \phi \text{ is an } A\text{-homomorphism}\}.$$

This is a subalgebra of $\text{End}_k(M)$. More generally, if M and N are A-modules we set

$$\text{Hom}_A(M, N) = \{\phi : M \longrightarrow N \mid \phi \text{ is an } A\text{-homomorphism}\}.$$

Arguing as in the proof of Lemma 10 above, we obtain:

Lemma 12 (Schur). *If k is algebraically closed and S and T are finite-dimensional simple A-modules, then*

$$\text{Hom}_A(S, T) \cong \begin{cases} k & \text{if } S \cong T \\ 0 & \text{otherwise.} \end{cases}$$

We can now give a complete classification of the finite-dimensional semisimple algebras.

Theorem 3 (Artin–Wedderburn). *Let A be a finite-dimensional algebra over an algebraically closed field k. Then A is semisimple if and only if*

$$A \cong M_{n_1}(k) \oplus M_{n_2}(k) \oplus \cdots \oplus M_{n_t}(k)$$

for some $t \in \mathbb{N}$ and $n_1, \ldots, n_t \in \mathbb{N}$.

Proof. (Sketch) We saw in Lemma 9 that $M_n(k)$ is a semisimple algebra, and if A and B are semisimple algebras, then it is easy to verify that $A \oplus B$ is semisimple.

For the reverse implication suppose that M and N are A-modules, with $M = \oplus_{i=1}^n M_i$ and $N = \oplus_{i=1}^m N_i$. The first claim is that $\text{Hom}_A(M, N)$ can be identified with the space of matrices

$$\{(\phi_{ij})_{1 \leq i \leq n, 1 \leq j \leq m} \mid \phi_{i,j} : M_j \longrightarrow N_i \text{ an } A\text{-homomorphism}\}$$

and that if $M = N$ with $M_i = N_i$ for all i then this space of matrices is an algebra by matrix multiplication, isomorphic to $\text{End}_A(M)$. This follows by an elementary calculation.

Now apply this to the special case where $M = N = A$, and

$$A = (S_1 \oplus S_2 \oplus \cdots \oplus S_{n_1}) \oplus (S_{n_1+1} \oplus \cdots \oplus S_{n_1+n_2}) \oplus \cdots$$
$$\oplus (S_{n_1+n_2+\cdots+n_{t-1}+1} \oplus \cdots \oplus S_{n_1+n_2+\cdots+n_t})$$

is a decomposition of A into simples such that two simples are isomorphic if and only if they occur in the same bracketed term. By Schur's Lemma above we see that ϕ_{ij} in this special case is 0 if S_i and S_j are in different bracketed terms, and is some $\lambda_{ij} \in k$ otherwise. There is then an obvious isomorphism of $\operatorname{Hom}_A(A, A)$ with $M_{n_1}(k) \oplus \cdots \oplus M_{n_t}(k)$. Finally, we note that for any algebra A, we have

$$\operatorname{End}_A(A, A) \cong A^{op}$$

and hence

$$A = (A^{op})^{op} \cong M_{n_1}(k)^{op} \oplus \cdots \oplus M_{n_t}(k)^{op}.$$

But it is easy to see that $M_n(k) \cong M_n(k)^{op}$ via the map taking a matrix X to its transpose, and so we are done. $\qquad\square$

We can also describe all the simple modules for such an algebra.

Corollary 1. *Suppose that*

$$A \cong M_{n_1}(k) \oplus M_{n_2}(k) \oplus \cdots \oplus M_{n_t}(k).$$

Then A has exactly t isomorphism classes of simple modules, one for each matrix algebra. If S_i is the simple corresponding to $M_{n_i}(k)$ then $\dim S_i = n_i$ and S_i occurs precisely n_i times in a decomposition of A into simple modules.

Proof. (Sketch) Choose a basis for A such that for each element $a \in A$ the map $x \longmapsto ax$ is given by a block matrix

$$\begin{pmatrix} A_1 & 0 & 0 & \cdots & 0 \\ 0 & A_2 & 0 & \cdots & 0 \\ \vdots & & & & \vdots \\ 0 & \cdots & 0 & 0 & A_t \end{pmatrix},$$

where $A_i \in M_{n_i}(k)$. Then A is the direct sum of the spaces given by the columns of this matrix, each of dimension n_i. Arguing as in Lemma 9 we see that each of these column spaces is a simple A-module. Swapping rows in a given block gives isomorphic modules. Thus, there are at most t non-isomorphic simples in a decomposition of A (and hence by Proposition 3 at most t isomorphism classes). Two simples from different blocks cannot be isomorphic (by considering the action of the matrix which is the identity in block A_i and zero elsewhere). $\qquad\square$

Remark 1. If k is not algebraically closed then the proofs of Lemmas 11 and 12 no longer hold. Instead, one deduces that for a simple module S the space $\operatorname{End}_A(S, S)$ is a division ring over k. (A *division ring* is a non-commutative version of a field.) There is then a version of the Artin–Wedderburn theorem, but where each $M_n(k)$ is replaced by some $M_n(D_i)$ with D_i some division ring containing k.

2.3.   *The Jacobson radical*

Suppose that A is not a semisimple algebra. One way to measure how far from semisimple it is would be to find an ideal I in A such that A/I is semisimple and I is minimal with this property.

Definition 20. The *Jacobson radical* (or just *radical*) of an algebra A, denoted $\mathcal{J}(A)$ (or just $\mathcal{J}$), is the set of elements $a \in A$ such that $aS = 0$ for all simple modules S. It is easy to verify that this is an ideal in A.

Definition 21. An ideal is *nilpotent* if there exists n such that $I^n = 0$. A *maximal submodule* in a module M is a module $L \subset M$ which is maximal by inclusion. The *annihilator* $\mathrm{Ann}(M)$ of a module M is the set of $a \in A$ such that $aM = 0$. This is easily seen to be a submodule of A.

When discussing the Jacobson radical, the following result is useful.

Lemma 13. *Let A be a finite-dimensional algebra. Then A has a largest nilpotent ideal.*

Proof. Consider the set of nilpotent ideals in A, and chose one, I, of maximal dimension. If J is another nilpotent ideal then the ideal $I + J$ is also nilpotent. (If $I^n = 0$ and $J^m = 0$ then $(I + J)^{m+n} = 0$, as the expansion of any expression $(a + b)^{n+m}$ with $a \in I$ and $b \in J$ contains at least n copies of a or m copies of b.) But then $\dim(I + J) = \dim I$ and hence $J \subseteq I$. $\qquad\square$

Theorem 4 (Jacobson). *Let A be a finite-dimensional algebra. The ideal $\mathcal{J}(A)$ is:*
(a) *The largest nilpotent ideal N in A.*
(b) *The intersection D of all maximal submodules of A.*
(c) *The smallest submodule R of A such that A/R is semisimple.*

Proof. (a) First, suppose that S is simple. Then NS is a submodule of S. If $NS = S$ then by induction $N^m S = S$ for all $m \geq 1$. But this contradicts the nilpotency of N, and so $N \subseteq \mathcal{J}$. For the reverse inclusion, consider a composition series for A

$$0 = A_n \subset A_{n-1} \subset \cdots \subset A_0 = A.$$

As A_i/A_{i+1} is simple we have $a(A_i/A_{i+1}) = 0$ for all $a \in \mathcal{J}$. But this implies that $\mathcal{J}A_i \subseteq A_{i+1}$, and hence

$$\mathcal{J}^n \subset \mathcal{J}^n A \subset A_n = 0.$$

(b) Suppose that $a \in \mathcal{J}$ and M is a maximal submodule of A. Then A/M is simple and so $a(A/M) = 0$. In other words, $a(1 + M) = 0 + M$ and so $a \in M$. Thus, $\mathcal{J} \subset M$ for every maximal submodule of A.

For the reverse inclusion, suppose that $\mathcal{J} \not\subseteq D$. Then there exists some simple S and $s \in S$ with $Ds \neq 0$. Now Ds is a submodule of S, and hence $Ds = S$. Thus, there exists $d \in D$ with $ds = s$; so $d - 1 \in \mathrm{Ann}(S) \not\subseteq A$, and there exists a maximal submodule M of A with $\mathrm{Ann}(S) \subseteq M$. But then $d \in D \subseteq M$ and $1 - d \in M$ implies that $1 \in M$, which contradicts $M \subset A$.

(c) (Sketch) First, we claim that D can be expressed as the intersection of finitely many maximal submodules of A. To see this, pick some submodule L which is the intersection of finitely many maximal submodules, such that $\dim L$ is minimal. Clearly, $D \subseteq L$. For any maximal M in A, we must have that $L = L \cap M$, and hence $L \subseteq D$.

Thus, $D = M_1 \cap M_2 \cap \cdots \cap M_n$ for some maximal submodules $M_1, \ldots M_n$. There is a homomorphism

$$\phi\colon \frac{A}{D} \longrightarrow \frac{A}{M_1} \oplus \cdots \frac{A}{M_n}$$

given by $\phi(a) = (a + M_1, \ldots, a + M_n)$. It is easy to see this is injective. As each M_i is maximal, we have embedded A/D into a semisimple module, and hence A/D is semisimple by Lemma 8.

Now suppose that A/X is semisimple. It remains to show that $D \subseteq X$. Write A/X as a direct sum of simples $S_i = L_i/X$. Then it is easy to check that the submodule $M_i = \sum_{i \neq j} L_i$ is a maximal submodule of A, and that the intersection of the M_i equals X. By definition, this intersection contains D, as required. $\qquad\square$

The Jacobson radical can be used to understand the structure of A-modules:

Lemma 14 (Nakayama). *If M is a finite-dimensional A-module such that $\mathcal{J}M = M$ then $M = 0$.*

Proof. (Sketch, for the case A is finite-dimensional) Suppose that $M \neq 0$ and choose a minimal set of generators $m_1, \ldots, m_t$ of M as an A-module. Now $m_t \in M = \mathcal{J}M$ implies that

$$m_t = \sum_{i=1}^{t} a_i m_i$$

for some $a_i \in \mathcal{J}$, and so

$$(1 - a_t)m_t = \sum_{i=1}^{t-1} a_i m_i.$$

Now $a_t \in \mathcal{J}$ implies that a_t is nilpotent, and then it is easy to check that $1 - a_t$ must be invertible. But this implies that m_t can be expressed in terms of the remaining m_i, which contradicts minimality.   □

We have the following generalisation of Nakayama's Lemma.

Proposition 4. *If A is a finite-dimensional algebra and M is a finite-dimensional A-module then $\mathcal{J}M$ equals:*
(a) The intersection D of all maximal submodules of M.
(b) The smallest submodule R of M such that M/R is semisimple.

Proof. (Sketch) Suppose that M_i is a maximal submodule of M. Then M/M_i is simple, and hence by Nakayama's Lemma $\mathcal{J}(M/M_i) = 0$. Therefore, $\mathcal{J}M \subseteq \mathcal{J}M_i \subseteq M_i$ and so $\mathcal{J}M \subseteq D$.

By Theorem 4, the module $M/\mathcal{J}M$ is semisimple, as it is a module for $A/\mathcal{J}$. Now suppose that L is a submodule of M such that M/L is semisimple. Let $M/L = M_1/L \oplus \cdots \oplus M_t/L$ where each M_i/L is simple. Then the modules $N_j = \sum_{i \neq j} M_i$ are maximal submodules of M and L is the intersection of the N_j. Hence $\mathcal{J}M$ is a submodule of L as $\mathcal{J}M$ is a submodule of D. Taking $L = \mathcal{J}M$, we see that D is a submodule of $\mathcal{J}M$ which completes the proof.   □

Motivated by the last result, we have:

Definition 22. The *radical* of a module M is defined to be the module $\mathcal{J}M$. Note that when $M = A$ this agrees with the earlier definition of the radical of an algebra. The *head* or *top* of M, denoted hd(M) or top(M), is the module $M/\mathcal{J}M$. By the last proposition the sequence

$$M \supset \mathcal{J}M \supset \mathcal{J}^2M \supset \cdots \supset \mathcal{J}^tM \supset \mathcal{J}^{t+1}M = 0$$

is such that each successive quotient is the largest semisimple quotient possible. This is called the *Loewy series* for M, and $t + 1$ is the *Loewy length* of M.

The head of a module M is the largest semisimple quotient of M. It can be shown that the submodule of M generated by all simple submodules is the largest semisimple submodule of M; we call this the *socle* of M, and denote it by soc(M).

2.4. The Krull–Schmidt theorem

Given a finite-dimensional A-module M, it is clear that we can decompose M as a direct sum of indecomposable modules. The Krull–Schmidt theorem says that this decomposition is essentially unique, and so it is enough to classify the indecomposable modules for an algebra.

Theorem 5 (Krull–Schmidt). *Let A be a finite-dimensional algebra and M be a finite-dimensional A-module. If*

$$M = M_1 \oplus M_2 \oplus \cdots \oplus M_n = N_1 \oplus N_2 \oplus \cdots \oplus N_m$$

are two decompositions of M into indecomposables then $n = m$ and there exists a permutation σ of $\{1, \ldots n\}$ such that $N_i \cong M_{\sigma(i)}$.

Proof. The idea is to proceed by induction on n, at each stage cancelling out summands which are known to be isomorphic. The details are slightly technical, and so will be omitted here. Instead, we will review below some of the ideas used in the proof. □

A key idea in the proof of the Krull–Schmidt theorem is the notion of a local algebra.

Definition 23. An algebra A is *local* if it has a unique maximal right (or left) ideal.

There are various characterisations of a local algebra.

Lemma 15. *Suppose that A is a finite-dimensional algebra over an algebraically closed field. Then the following are equivalent:*
(a) *A is a local algebra.*
(b) *The set of non-invertible elements of A form an ideal.*
(c) *The only idempotents in A are 0 and 1.*
(d) *The quotient $A/\mathcal{J}$ is isomorphic to k.*

Proof. This is not difficult, but is omitted as it requires a few preparatory results. □

Example 12. It is not hard to show that $k[x]/(x^n)$ is a local algebra. With a little more work it can be shown that kG is a local algebra when G is of order p^n and k has characteristic p.

Remark 2. In fact, (a) and (b) are equivalent for any algebra A. However, there exist examples of infinite dimensional algebras with only 0 and 1 as idempotents which are not local, for example $k[x, y]$. Also, if the field is not algebraically closed then $A/\mathcal{J}$ will only be a division ring in general.

Lemma 16 (Fitting). *Let M be a finite-dimensional A-module, and $\phi \in \mathrm{End}_A(M)$. Then for large enough n we have*

$$M = \mathrm{im}(\phi^n) \oplus \ker(\phi^n).$$

In particular, if M is indecomposable then any non-invertible endomorphism of M must be nilpotent.

Proof. Note that $\phi^{i+1}(M) \subseteq \phi^i(M)$ for all i. As M is finite-dimensional, there must exist an n such that $\phi^{n+t}(M) = \phi^n(M)$, for all $t \geq 1$ and so ϕ^n is an isomorphism from $\phi^n(M)$ to $\phi^{2n}(M)$. For $m \in M$, let x be an element such that $\phi^n(m) = \phi^{2n}(x)$. Now

$$m = \phi^n(x) + (m - \phi^n(x)) \in \mathrm{im}(\phi^n) + \ker(\phi^n)$$

and so $M = \mathrm{im}(\phi^n) + \ker(\phi^n)$. If $\phi^n(m) \in \mathrm{im}(\phi^n) \cap \ker(\phi^n)$ then $\phi^{2n}(m) = 0$, and so $\phi^n(m) = 0$. Thus the sum is direct, as required.   □

Local algebras are useful as they allow us to detect indecomposable modules.

Lemma 17. *Let M be a finite-dimensional A-module. Then M is indecomposable if and only if $\mathrm{End}_A(M)$ is a local algebra.*

Proof. First, suppose that $M = M_1 \oplus M_2$, and for $i = 1, 2$ let e_i be the map from M to M which maps $m_1 + m_2$ to m_i. Then $e_i \in \mathrm{End}_A(M)$ is non-invertible (as it has non-zero kernel). But $e_1 + e_2 = 1$, which is invertible, which implies that $\mathrm{End}_A(M)$ is not local by Lemma 15.

Now suppose that M is indecomposable. Let I be a maximal right ideal in $\mathrm{End}_A(M)$, and pick $\phi \in \mathrm{End}_A(M) \backslash I$. By maximality, we have $\mathrm{End}_A(M) = \mathrm{End}_A(M)\phi + I$. Thus, we can write $1 = \theta\phi + \mu$ where $\theta \in \mathrm{End}_A(M)$ and $\mu \in I$. Note that any element in I cannot be an isomorphism of M (as it would then be invertible), and hence by Fitting's Lemma we have that $\mu^n = 0$ for some $n >> 0$. But then

$$(1 + \mu + \mu^2 + \cdots + \mu^{n-1})\theta\phi = (1 + \mu + \mu^2 + \cdots + \mu^{n-1})(1 - \mu) = 1 - \mu^n = 1$$

and so ϕ is an isomorphism. But then I consists precisely of the non-invertible elements in $\mathrm{End}_A(M)$, and the result follows by Lemma 15.   □

3. Projective and Injective Modules

3.1. *Projective and injective modules*

Definition 24. A *short exact sequence* of A-modules is a sequence of the form

$$0 \longrightarrow L \xrightarrow{\phi} M \xrightarrow{\psi} N \to 0,$$

such that the map ϕ is injective, the map ψ is surjective, and $\operatorname{im} \phi = \ker \psi$. More generally, a sequence

$$\cdots \longrightarrow L \xrightarrow{\phi} M \xrightarrow{\psi} N \longrightarrow \cdots$$

is *exact at M* if $\operatorname{im} \phi = \ker \psi$. If a sequence is exact at every module then it is called *exact*. (Thus a short exact sequence is exact.)

Note that in a short exact sequence as above we have that

$$M/L \cong N,$$

by the isomorphism theorem, and $\dim M = \dim L + \dim N$. When a sequence starts or ends in a 0 it is common to assume that it is exact (as we will do in what follows).

Lemma 18. *Given a short exact sequence of A-modules*

$$0 \longrightarrow L \xrightarrow{\phi} M \xrightarrow{\psi} N \to 0,$$

the following are equivalent:
(a) *There exists a homomorphism $\theta : N \longrightarrow M$ such that $\psi\theta = id_N$.*
(b) *There exists a homomorphism $\tau : M \longrightarrow L$ such that $\tau\phi = id_L$.*
(c) *There is a module X with $M = X \oplus \ker(\psi)$.*

Proof. (Sketch) We will show that (a) is equivalent to (c); that (b) is equivalent to (c) is similar. First, suppose that θ as in (a) exists. Then θ must be an injective map. Let $X = \operatorname{im}(\theta)$, a submodule of M isomorphic to N. It is easy to check that $X \cap \ker(\psi) = 0$ and that $\dim(X \oplus \ker(\psi)) = \dim M$ by exactness at M. Therefore, $M = X \oplus \ker(\psi)$.

Now suppose that $M = X \oplus \ker(\psi)$. Consider the restriction of ψ to X; it is clearly an isomorphism and so θ can be taken to be an inverse to ψ. $\square$

Definition 25. An A-module P is *projective* if for all surjective A-module homomorphisms $\theta : M \longrightarrow N$ and for all $\phi : P \longrightarrow N$ there exists $\psi : P \longrightarrow M$ such that $\theta\psi = \phi$.

Thus, a module P is projective if there always exists ψ such that the following diagram commutes

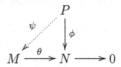

(Note that here we are using our convention about exactness for the bottom row in the diagram.)

There is a dual definition, obtained by reversing all the arrows and swapping surjective and injective.

Definition 26. An A-module I is *injective* if for all injective A-module homomorphisms $\theta : N \longrightarrow M$ and for all $\phi : N \longrightarrow I$ there exists $\psi : M \longrightarrow I$ such that $\psi\theta = \phi$.

Thus, a module I is injective if there always exists ψ such that the following diagram commutes

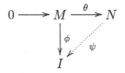

Example 13. For $m \geq 1$, the module A^m is projective. To see this, denote the ith coordinate vector $(0, \ldots, 0, 1, 0 \ldots, 0)$ by v_i, and suppose that $\phi(v_i) = n_i \in N$. As g is surjective, there exists $m_i \in M$ such that $g(m_i) = n_i$. Given a general element $(a_1, \ldots, a_m) \in A^m$, define

$$\psi(a_1, \ldots, a_m) = \sum_{i=1}^{m} a_i m_i.$$

It is easy to verify that this gives the desired A-module homomorphism.

We would like a means to recognise projective modules P without having to consider all possible surjections and morphisms from P. The following lemma provides this, and shows that the above example is typical.

Lemma 19. *For an algebra A, the following are equivalent.*
(a) *P is projective.*
(b) *Whenever $\theta : M \longrightarrow P$ is a surjection then $M \cong P \oplus \ker(\theta)$.*
(c) *P is isomorphic to a direct summand of A^m for some m.*

Proof. First, suppose that P is projective. We have by definition a commutative diagram

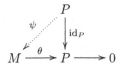

and by Lemma 18 this implies that $M \cong P \oplus \ker(\theta)$.

Now suppose that (b) holds. Given any A-module M with generators m_i, $i \in I$, there is a surjection from $\oplus_{i \in I} A$ onto M given by the map taking 1 in the ith copy of A to m_i. Taking $M = P$ a projective we deduce that (c) holds.

Finally suppose that $A^m \cong P \oplus X$ for some P and X. Let π be the projection map from A^n onto P, and ι be the inclusion map from P into A^m. Given modules M and N and a surjection from M to N, we have the commutative diagram

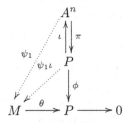

It is easy to check that $\psi_1 \iota$ gives the desired map ψ for P in the definition of a projective module. $\qquad\square$

Suppose that A is finite-dimensional and

$$A = P(1) \oplus \cdots \oplus P(n) \tag{1}$$

is a decomposition of A into indecomposable direct summands. By the last result, these summands are indecomposable projective modules.

Lemma 20. *Suppose that A is a finite-dimensional algebra. Let P be a projective A-module with submodule N, and suppose that every homomorphism $\phi : P \longrightarrow P$ maps N to N. Then there is a surjection from $\mathrm{End}_A(P)$ onto $\mathrm{End}_A(P/N)$ and if P is indecomposable then so is P/N.*

Proof. (Sketch) Given $\phi : P \longrightarrow P$ let $\bar{\phi}$ be the obvious map from P/N to P/N. Check this is well-defined; it is clearly a homomorphism. The map $\phi \longrightarrow \bar{\phi}$ gives an algebra homomorphism from $\mathrm{End}_A(P)$ to $\mathrm{End}_A(P/N)$;

given $\psi \in \mathrm{End}_A(P/N)$ use the projective property of P to construct a map ϕ so that $\bar{\phi} = \psi$.

If P is indecomposable then $\mathrm{End}_A(P)$ is a local algebra by Lemma 17. Therefore, there is a unique maximal right ideal in $\mathrm{End}_A(P)$, and hence a unique maximal ideal in $\mathrm{End}_A(P/N)$ (as we have shown that this is a quotient of $\mathrm{End}_A(P)$). Thus, $\mathrm{End}_A(P/N)$ is local, and hence P/N is indecomposable. $\square$

Theorem 6. *Let A be a finite-dimensional algebra, and decompose A as in (1). Setting $S(i) = P(i)/\mathcal{J}P(i)$, we have:*
(a) *The module $S(i)$ is simple, and every simple A-module is isomorphic to some $S(i)$.*
(b) *We have $S(i) \cong S(j)$ if and only if $P(i) \cong P(j)$.*

Proof. (Sketch) (a) The modules $S(i)$ is semisimple, so it is enough to check it is indecomposable. Note that $P(i)$ and $\mathcal{J}P(i)$ satisfy the assumptions in Lemma 20, and so $S(i)$ is indecomposable.

Let S be a simple module and choose $x \neq 0$ in S. As $1x = x$ there is some $P(i)$ such that $P(i)x \neq 0$ (as $Ax \neq 0$ and A is the direct sum of the $P(i)$). Define a homomorphism from $P(i)$ to S, so S is a simple quotient of $P(i)$. But $\mathcal{J}P(i)$ is the unique maximal submodule of $P(i)$, and so as S is simple we have $S \cong P(i)/\mathcal{J}P(i)$.

(b) If $P(i) \cong P(j)$ via ϕ it is easy to see that $\phi(\mathcal{J}P(i)) \subseteq \mathcal{J}P(j)$). Hence, ϕ induces a homomorphism from $S(i)$ to S_j. As ϕ is invertible, this has an inverse, and so $S(i) \cong S(j)$ by Schur's Lemma.

If $S(i) \cong S(j)$, then use the projective property to construct a homomorphism ψ from $P(i)$ to $P(j)$. Show that the image of this map cannot be inside $\mathcal{J}P(j)$, so as $\mathcal{J}P(j)$ is a maximal submodule ψ must have image all of $P(j)$. By Lemma 19, we deduce that $P(i) \cong P(j) \oplus \ker(\psi)$, and so as $P(i)$ is indecomposable we have $\ker(\psi) = 0$. Thus ψ is an isomorphism. $\square$

By Krull–Schmidt, this implies that a finite-dimensional algebra A has only finitely many isomorphism classes of simple modules.

Definition 27. Let M be a finite-dimensional A-module. A *projective cover* for M is a projective module P such that

$$P/\mathcal{J}P \cong M/\mathcal{J}M$$

and there exists a surjection $\pi : P \longrightarrow M$.

Lemma 21. *Let A be a finite-dimensional algebra. Every finite-dimensional A-module has a projective cover, which is unique up to isomorphism. In particular, suppose that*

$$M/\mathcal{J}M \cong S(1)^{n_1} \oplus S(2)^{n_2} \oplus \cdots \oplus S(t)^{n_t}.$$

Then

$$P = P(1)^{n_1} \oplus P(2)^{n_2} \oplus \cdots \oplus P(t)^{n_t}$$

is a projective cover of M via the canonical surjection on each component.

Proof. (Sketch) It is clear that the given P satisfies $P/\mathcal{J}P \cong M/\mathcal{J}M$. Use the projective property to construct a homomorphism π from P to M; it is easy to see that $\mathrm{im}(\pi) + \mathcal{J}M = M$ by the commutativity of the related diagram. But then $\mathcal{J}(M/\mathrm{im}(\pi)) = (\mathcal{J}M + \mathrm{im}(\pi))/\mathrm{im}(\pi) = M/\mathrm{im}(\pi)$ and so by Nakayama's Lemma we have $M/N = 0$. Thus π is surjective as required. □

Definition 28. A *projective resolution* of a module M is an exact sequence

$$\cdots \longrightarrow P_3 \longrightarrow P_2 \longrightarrow P_1 \longrightarrow M \longrightarrow 0$$

such that all the P_i are projective.

By induction using Lemma 21 we deduce:

Proposition 5. *If A is a finite-dimensional algebra then every finite-dimensional A-module has a projective resolution.*

There is a similar theory for injective modules, but instead of developing this separately we will instead use dual modules to relate the two.

The injective analogue of a projective cover is called the *injective envelope* of M. An *injective resolution* of M is an exact sequence

$$0 \longrightarrow M \longrightarrow I_1 \longrightarrow I_2 \longrightarrow I_3 \longrightarrow \cdots$$

such that all the I_i are injective.

Theorem 7. *Suppose that A is a finite-dimensional algebra and M a finite-dimensional A-module.*
(a) *M is simple if and only if M^* is a simple A^{op}-module.*
(b) *M is projective if and only if M^* is an injective A^{op}-module.*
(c) *M is injective if and only if M^* is a projective A^{op}-module.*
(d) *The injective envelope of M is I if and only if the projective cover of the A^{op}-module M^* is I^*.*

Proof. This is a straightforward application of duality.   □

Projective and injective modules play a crucial role in the study of the cohomology of representations. In a non-semisimple representation theory, there are certain spaces associated to $\mathrm{Hom}_A(M, N)$ called *extension groups* $\mathrm{Ext}^i_A(M, N)$. To introduce these properly, we would need to work with the category of modules, and introduce the notion of a derived functor. Unfortunately this is beyond the scope of the current chapter.

3.2.   *Idempotents and direct sum decompositions*

Every algebra has at least two idempotents, 0 and 1. If A is not local then there exists another idempotent $e \in A$ and e and $1 - e$ are two non-zero orthogonal idempotents, giving a decomposition of A-modules

$$A = Ae \oplus A(1 - e).$$

If e is a central idempotent then so is $1 - e$, and the above decomposition becomes a direct sum of algebras. Conversely, if $A = M_1 \oplus M_2$ as an A-module, then the corresponding decomposition $1 = e_1 + e_2$ is an orthogonal idempotent decomposition of 1. If the decomposition of A is as a direct sum of algebras, then the corresponding idempotents are central.

Definition 29. We say that an algebra is *connected* or *indecomposable* if 0 and 1 are the only central idempotents in A.

Note that if A is not connected, say $A = A_1 \oplus A_2$, then any A-module M decomposes as a direct sum $M_1 \oplus M_2$ where M_i is an A_i-module for $i = 1, 2$. (This follows by decomposing $1 \in A$ and applying it to M.) Thus, we can reduce the study of the representations of an algebra to the case where the algebra is connected.

Suppose that A is a finite-dimensional algebra. By repeatedly decomposing A as an A-module, we can write

$$A = P_1 \oplus \cdots \oplus P_n,$$

where the P_i are indecomposable left ideals in A. (The sum is finite as A is finite-dimensional.) There is a corresponding decomposition of 1 as a sum of primitive orthogonal idempotents. Conversely, any such decomposition of 1 gives rise to a decomposition of A into indecomposable left ideals. Note that we can identify primitive idempotents by the following application of Lemma 17:

Corollary 2. *Suppose that A is a finite-dimensional algebra. Then an idempotent $e \in A$ is primitive if and only if eAe is local.*

Definition 30. Suppose that A is a finite-dimensional algebra with a complete set $\{e_1, \ldots e_n\}$ of primitive orthogonal idempotents. Then A is *basic* if $Ae_i \cong Ae_j$ implies that $i = j$.

Basic algebras have the following nice properties.

Proposition 6. *Suppose that k is algebraically closed.*
(a) *A finite-dimensional k-algebra A is basic if and only if*

$$A/\mathcal{J} \cong k \times k \times \cdots \times k.$$

(b) *Every simple module over a basic algebra is one dimensional.*

Proof. (Sketch) (a) Suppose that A is basic and consider a complete set of primitive idempotents $e_1, \ldots, e_n$ for A. By Theorem 6 the modules $S_i = (A/\mathcal{J})e_i$ are simple $A/\mathcal{J}$-modules. Also, as A is basic these simples are non-isomorphic. Then Schur's Lemma implies that Hom-spaces between such simples are isomorphic to 0 or k, and one can define an injective homomorphism

$$\phi : A/\mathcal{J} \longrightarrow \mathrm{End}_{A/\mathcal{J}}(S_1 \oplus \cdots \oplus S_n) \cong k \times \cdots \times k.$$

By dimensions this is an isomorphism.

If $A/\mathcal{J}$ is basic then the S_i above are all non-isomorphic (as the primitive idempotents are even central in $A/\mathcal{J}$), and the same argument as above implies that A is basic.

(b) Any simple A-module is also an $A/\mathcal{J}$-module by Nakayama's Lemma. But by part (a) this is isomorphic to $k \times \cdots \times k$, which implies the result. □

Given an arbitrary finite-dimensional algebra A, we can associate a basic algebra to it in the following manner.

Definition 31. Suppose that A is finite-dimensional and has a complete set $\{e_1, \ldots e_n\}$ of primitive orthogonal idempotents. Pick idempotents $e_{i_1}, \ldots e_{i_t}$ from this set such that $Ae_{i_a} \cong Ae_{i_b}$ implies that $a = b$, and so that the collection is maximal with this property. Then define

$$e_A = \sum_{a=1}^{t} e_{i_a}$$

and set $A^b = e_A A e_A$, the *basic algebra associated to* A. (It is easy to see that this is indeed a basic algebra, and is independent of the choice of idempotents.)

As we have not given a precise definition of a category, we shall state the next result without proof.

Theorem 8. *Suppose that A is a finite-dimensional algebra. Then the category of finite-dimensional A-modules is equivalent to the category of finite-dimensional A^b-modules.*

This means that to understand the representation theory of a finite-dimensional algebra it is enough to consider representations of the corresponding basic algebra.

Let us now consider the special case of the path algebra of a quiver.

Lemma 22. *Let Q be a finite quiver. Then the sum*

$$1 = \sum_{i \in Q_0} \epsilon_i$$

is a decomposition into a complete set of primitive orthogonal idempotents for kQ.

Proof. All that remains to prove is that the ϵ_i are primitive, for which it is enough to show that $\epsilon_i k Q \epsilon_i$ is local. Suppose that $e \in \epsilon_i k Q \epsilon_i$ is an idempotent. Then $e = \lambda \epsilon_i + w$ where $\lambda \in k$ and w is a sum of paths from a to a. But then

$$0 = e^2 - e = (\lambda^2 - \lambda)\epsilon_i + w^2 + (2\lambda - 1)w$$

implies that $w = 0$ and $\lambda = 0$ or $\lambda = 1$. $\square$

We can now characterise the connected path algebras of quivers.

Lemma 23. *Let Q be a finite quiver. Then kQ is connected if and only if Q is a connected quiver.*

Proof. (Sketch) It can be shown that kQ is connected if and only if there does not exist a partition $Q_0 = X \cup Y$ of the set of vertices such that for all $x \in X$ and $y \in Y$ we have $\epsilon_x kQ\epsilon_y = 0 = \epsilon_y kQ\epsilon_x$.

Clearly, if Q is not connected then there exists a partition $Q_0 = X \cup Y$ so that there is no path from a vertex in X to a vertex in Y (or vice versa). Thus in this case $\epsilon_x kQ\epsilon_y = 0 = \epsilon_y kQ\epsilon_x$, and kQ is not connected.

If kQ is not connected but Q is connected, there exists a partition $Q_0 = X \cup Y$ as above, and elements $x \in X$ and $y \in Y$ with an arrow α: $x \longrightarrow y$ in Q. But then $\alpha \in \epsilon_x kQ\epsilon_y$ which contradicts our assumption on kQ. □

Theorem 9. *Let Q be a finite, connected, acyclic quiver. Then kQ is a basic connected algebra with radical given by the arrow ideal.*

Proof. (Sketch) By Lemma 22, we have a decomposition

$$kQ/R_Q = \bigoplus_{a,b \in Q_o} \epsilon_a(kQ/R_Q)\epsilon_b.$$

As R contains all non-trivial paths each summand is non-zero only when $a = b$, in which case it is isomorphic to k. Thus, we will be done by Proposition 6 and Lemma 23 if we can show that $R_Q = \mathcal{J}$. But as Q is acyclic there exists a maximal path length in Q. Hence, $R_Q^n = 0$ for $n \gg 0$, and so $R \subseteq \mathcal{J}$ by Theorem 4. It is not too hard to show that in fact any nilpotent ideal I such that A/I is a product of copies of k must equal $\mathcal{J}(A)$. □

Now we consider the case of bound quiver algebras.

Proposition 7. *Let Q be a finite quiver with admissible ideal I in kQ. Then*
(a) The set

$$\{e_i = \epsilon_i + I : i \in Q_0\}$$

is a complete set of primitive orthogonal idempotents in kQ/I.
(b) The algebra kQ/I is connected if and only if Q is a connected quiver.
(c) The algebra kQ/I is basic, with radical R_Q/I.

Proof. (Sketch) The proofs of (a) and (b) are similar to those for kQ. Part (c) is almost immediate from the corresponding result for kQ. □

We have seen that the representation theory of finite-dimensional algebras reduces to the study of connected basic algebras. The last result says that bound quiver algebras for connected quivers are such algebras. We conclude this section with the following theorem.

Theorem 10. *Let A be a basic, connected, finite-dimensional k-algebra over an algebraically closed field. Then there is a connected quiver Q associated to A and an admissible ideal I in kQ such that*

$$A \cong kQ/I.$$

Thus, over algebraically closed fields the study of finite-dimensional algebras can be reduced to the study of bound quiver algebras.

Proof. (Sketch) Rather than give a detailed proof, we will sketch how to construct the quiver associated to A.

Let $\{e_1, \ldots, e_n\}$ be a complete set of primitive orthogonal idempotents in A. Then Q has vertex set $\{1, \ldots, n\}$. Given $1 \leq i, j \leq n$, the number of arrows from i to j equals the dimension of the vector space $e_i(\mathcal{J}/\mathcal{J}^2)e_j$.

One then checks that this quiver is independent of the choice of idempotents and is connected. Then one defines a homomorphism from kQ to A, and shows that this is (i) surjective and (ii) has kernel which is an admissible ideal in kQ. The result then follows from the first isomorphism theorem.

$\square$

For a discussion of what happens when k is not algebraically closed, see [3, Section 4.1].

3.3. *Simple and projective modules for bound quiver algebras*

In general, it is hard to determine explicitly the simple modules for an algebra. Indeed, some of the most important open questions in representation theory relate to determining simple modules. However, in the case of a bound quiver algebra the simple modules can be written down entirely explicitly.

We will also see that the indecomposable projectives can also be easily constructed. The same is true for indecomposable injectives, but we will not consider these in detail here.

Let kQ/I be a bound quiver algebra. We know by Proposition 7 and Theorem 6 that the simple modules are parameterised by the vertices of Q, and are all one dimensional (as the algebras are basic). Given this, the following result is almost clear.

Proposition 8. *Let kQ/I be a bound quiver algebra. For $a \in Q_0$, let $S(a)$ be the representation of Q such that*

$$S(a)_b = \begin{cases} k & a = b \\ 0 & a \neq b \end{cases}$$

and for all arrows α the map $\phi_\alpha = 0$. Then

$$\{S(a) : a \in Q_0\}$$

is a complete set of non-isomorphic simple modules for kQ/I.

Proof. The only thing that remains to be checked is that the various simples are not isomorphic, but this is straightforward. $\qquad\square$

The description of the projective modules $P(a)$ is slightly more complicated.

Proposition 9. *Let kQ/I be a bound quiver algebra, and $P(a)$ the projective corresponding to ϵ_a. Then $P(a)$ can be realised in the following manner. For $b \in Q_0$ let $P(a)_b$ be the k-vector space with basis the set of all elements of the form $w + I$ where w is a path from a to b. Given an arrow $\alpha : b \longrightarrow c$, the map $\phi_\alpha : P(a)_b \longrightarrow P(a)_c$ is given by left multiplication by $\alpha + I$.*

Proof. This is a straightforward consequence of the explicit identification of quiver representations with kQ/I-modules given earlier. $\qquad\square$

The description of injective modules for a bound quiver algebra is similar, using Theorem 7.

4.   Exercises

(1) Classify the simple modules for the cyclic group C_n over an algebraically closed field of characteristic $p \geq 0$.
(2) Suppose that Q is a finite quiver containing some cycle. Show that Q has infinitely many non-isomorphic simple representations over $\mathbb{C}$.
(3) Let $A = k[x]$ and M be the two-dimensional A module where x acts via the matrix

$$\begin{pmatrix} 0 & 1 \\ 0 & 0 \end{pmatrix}$$

with respect to some basis of M. Prove that M is not a semisimple module.

(4) Determine the indecomposable projectives and their radicals for the bound quiver

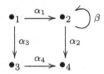

with the relations $\beta^2 = 0$ and $\alpha_2\alpha_1 = 2\alpha_4\alpha_3$.

(5) Classify the indecomposable representations of the quiver Q given by

$$\bullet_1 \xrightarrow{\alpha_1} \bullet_2 \xrightarrow{\alpha_2} \bullet_3 \xrightarrow{\alpha_3} \cdots \xrightarrow{\alpha_{n-2}} \bullet_{n-1} \xrightarrow{\alpha_{n-1}} \bullet_n \ .$$

5. Further Reading

For a far more comprehensive treatment of this material (together with many more examples and exercises), the reader is recommended to look at [1, Chapters I–III]. For simplicity, [1] only considers algebras over algebraically closed fields. An excellent (if rapid) introduction which considers more general rings can be found in [3, Chapters 1 and 4].

A very accessible introduction which integrates the representation theories of Lie algebras, finite groups and quivers, in a holistic manner can be found in [4]. Two books which go into more advanced topics than this chapter (and which are not entirely suitable for the beginner) are [2] and [5].

6. Outline Solutions to the Exercises

(1) There are various ways to approach this. One is to prove the following more general result first.

The *centre* of an algebra A, denoted $Z(A)$, is the set of $z \in A$ such that $za = az$ for all $a \in A$. This is a subalgebra of A. If k is algebraically closed and S is a simple A-module show that for all $z \in Z(A)$ there exists $\lambda \in k$ such that $zm = \lambda m$ for all $m \in S$.

This reduces the problem to classifying the one-dimensional representations of kC_n. Let $n = p^t m$ where m and p are coprime (or $n = m$ if $p = 0$). Show that for each primitive mth root of unity λ_i there is a one-dimensional representation where a generator x of C_n acts by λ_i, and no others can occur.

(2) For $\lambda \in \mathbb{C}\backslash\{0\}$, consider the representation $M(\lambda)$ of Q given by one-dimensional vector spaces at each vertex in the cycle, with all arrows in the cycle labelled by the identity map except one which corresponds to multiplication by λ. Show that $M(\lambda)$ is simple, using the (easy) fact that if a subrepresentation N of M has a non-zero vector space at a vertex i with an injective map from i to j in M then N has a non-zero vector space at j. It is easy to see that the $M(\lambda)$ as λ varies are pairwise non-isomorphic.

(3) If M is semisimple then there must be two distinct one-dimensional submodules, each spanned by a different linear combination of the two basis vectors e_1 and e_2. However, it is easy to see that the only linear combination generating a one-dimensional submodule of M is e_1 alone. More generally it is not hard to show that the finite-dimensional indecomposable modules for $k[x]$ correspond to the Jordan blocks $J_n(\lambda)$ with $n \in \mathbb{N}$ and eigenvalue $\lambda \in k$.

(4) An application of Theorem 6, Proposition 8 and Proposition 9.

(5) Let $M = (M_i, \phi_i)$ be an indecomposable representation of Q.

    (a) Show that if ϕ_i is not injective then $M_j = 0$ for $j > i$.

    (b) Similarly show that if ϕ_i is not surjective then $M_j = 0$ for $j \leq i$.

    (c) Deduce that M is isomorphic to a representation of the form

$$\cdots \longrightarrow 0 \longrightarrow k \xrightarrow{\text{id}} \cdots \xrightarrow{\text{id}} k \longrightarrow 0 \longrightarrow \cdots .$$

    (d) Show that the $\frac{n(n+1)}{2}$ such modules are pairwise non-isomorphic.

In general, it is provably impossible to classify the indecomposable modules of an algebra, let alone to expect to have only finitely many such. One of the highlights of the theory of finite-dimensional algebras is *Gabriel's Theorem*, which classifies those quivers with only finitely many indecomposable representations (in terms of Dynkin diagrams of types A, D, or E), and further shows that for such quivers the indecomposable representations correspond to positive roots for the associated Dynkin diagram. This example is a simple special case of that result. A very readable exposition of Gabriel's Theorem can be found in [4, Chapter 6], and in greater detail in [1, Chapter VII].

References

[1] I. Assem, D. Simson and A. Skowroński, *Elements of the representation theory of associative algebras I*, LMS student texts, **65**, Cambridge, 2006.

[2] M. Auslander, I. Reiten and S. Smalø, *Representation theory of Artin algebras*, Cambridge studies in advanced mathematics, **36**, Cambridge, 1994.

[3] D. J. Benson, *Representations and cohomology I*, Cambridge studies in advanced mathematics, **30**, Cambridge, 1991.

[4] P. Etingof, O. Goldberg, S. Hensel, T. Liu, A. Schwender, D. Vaintrob and E. Yudovina, *Introduction to representation theory*. AMS student mathematical library, **59**, AMS, 2011.

[5] P. Gabriel and A. V. Roiter, *Representations of finite-dimensional algebras*, Springer, 1997.

Chapter 4

The Invariant Theory of Finite Groups

Peter Fleischmann and James Shank

School of Mathematics, Statistics and Actuarial Science
University of Kent, Canterbury CT2 7NF, UK
P.Fleischmann@kent.ac.uk, R.J.Shank@kent.ac.uk

Mathematicians seek to exploit all available symmetry and often encode symmetry using the language of group actions. In this chapter, we consider finite groups acting by ring automorphisms on a polynomial ring. Our goal is to understand the subring of invariant polynomials.

1. Introduction

Invariant Theory is the study of polynomials with specified symmetry. Our starting point is a representation V of a group G over a field $\mathbf{F}$. Thus, G acts on V by linear transformations. Our convention is that V is a left module over the group ring $\mathbf{F}G$. The dual vector space $V^* := \hom_\mathbf{F}(V, \mathbf{F})$ is a natural right $\mathbf{F}G$ module with the action determined by $(\phi \cdot g)(v) = \phi(g \cdot v)$ for $g \in G$, $v \in V$ and $\phi \in V^*$. If the dimension of V as a vector space over $\mathbf{F}$ is n, then choosing a basis for V gives a group homomorphism $\rho : G \to \mathrm{GL}_n(\mathbf{F})$ and the action of $g \in G$ on V corresponds to the multiplication of a column vector on the left by the matrix $\rho(g)$ and, if we use the dual basis for V^*, the action on V^* corresponds to the multiplication of a row vector on the right by $\rho(g)$. It is sometimes convenient to convert the the right action of $\mathbf{F}G$ on V^* to a left action by defining $g \cdot \phi := \phi \cdot g^{-1}$.

Example 1. For any field $\mathbf{F}$, the symmetric group Σ_3 acts on $V := \mathbf{F}^3$ by permuting the standard basis. If we use this action to define the homomorphism $\rho : \Sigma_3 \to \mathrm{GL}_3(\mathbf{F})$, then for the three-cycle $(1\,2\,3)$, we have

$$\rho((1\,2\,3)) = \begin{pmatrix} 0 & 0 & 1 \\ 1 & 0 & 0 \\ 0 & 1 & 0 \end{pmatrix}.$$

If $\{x_1, x_2, x_3\}$ denotes the basis for V^* dual to the standard basis, then, since $[1\,0\,0]\rho((1\,2\,3)) = [0\,0\,1]$, we have $x_1 \cdot (1\,2\,3) = x_3$.

We use $\mathbf{F}[V]$ to denote the symmetric algebra on V^*. By choosing a basis $\{x_1, x_2, \ldots, x_n\}$ for V^*, we identify $\mathbf{F}[V]$ with the algebra of polynomials $\mathbf{F}[x_1, x_2, \ldots, x_n]$. The action of G on V^* extends to an action by degree preserving algebra automorphisms on $\mathbf{F}[V]$. The *ring of invariants*, $\mathbf{F}[V]^G$, is the subalgebra of $\mathbf{F}[V]$ consisting of those polynomials fixed by the group action. The elements of $\mathbf{F}[V]$ represent polynomial functions on V and the elements of $\mathbf{F}[V]^G$ represent polynomial functions on the space of orbits V/G. If G is finite and $\mathbf{F}$ is algebraically closed, $\mathbf{F}[V]^G$ is the ring of regular functions on the categorical quotient $V /\!\!/ G$ (see §2.3 of Derksen and Kemper [15] for a definition of the categorical quotient).

Example 2. As in Example 1, consider the permutation action of Σ_3 on $\mathbf{F}^3$. The first three elementary symmetric functions are

$$\sigma_1 := x_1 + x_2 + x_3, \quad \sigma_2 := x_1 x_2 + x_1 x_3 + x_2 x_3 \text{ and } \sigma_3 := x_1 x_2 x_3.$$

It is clear that $\sigma_i \in \mathbf{F}[V]^{\Sigma_3}$. It is not hard to show that every symmetric polynomial can be written as a polynomial in the elementary symmetric functions. Furthermore, $\{\sigma_1, \sigma_2, \sigma_3\}$ is algebraically independent. We summarise these two observations by writing $\mathbf{F}[V]^{\Sigma_3} = \mathbf{F}[\sigma_1, \sigma_2, \sigma_3]$. If we restrict to the subgroup $A_3 = \langle (1\,2\,3) \rangle$, then $\mathbf{F}[V]^{\Sigma_3} \subset \mathbf{F}[V]^{A_3}$. As long as the characteristic of $\mathbf{F}$ is not 2,

$$d := (x_1 - x_2)(x_1 - x_3)(x_2 - x_3) \in \mathbf{F}[V]^{A_3} \setminus \mathbf{F}[V]^{\Sigma_3}.$$

(If the characteristic of $\mathbf{F}$ is 2, we can take $d = x_1^2 x_2 + x_2^2 x_3 + x_3^2 x_1$.) It is not too hard to show that $d^2 \in \mathbf{F}[V]^{\Sigma_3}$ and that $\mathbf{F}[V]^{A_3}$ is the free $\mathbf{F}[V]^{\Sigma_3}$-module of rank 2 generated by $\{1, d\}$.

We assume throughout that G is finite. In this case, $\mathbf{F}[V]^G$ is a finitely generated algebra. Our goal is to understand $\mathbf{F}[V]^G$. We would like to be able to effectively compute generators for $\mathbf{F}[V]^G$ and to understand how properties of the representation V are reflected in properties of the ring $\mathbf{F}[V]^G$. If the the order of G is a unit in $\mathbf{F}$ then the *Reynolds operator*

$$\mathcal{R} := \frac{1}{|G|} \sum_{g \in G} g$$

induces a surjective $\mathbf{F}[V]^G$-homomorphism from $\mathbf{F}[V]$ to $\mathbf{F}[V]^G$, and provides a powerful tool to study $\mathbf{F}[V]^G$. However, if the characteristic of $\mathbf{F}$ is a prime p which divides the order of G, in other words, if V is a *modular representation*, then it is much harder to understand $\mathbf{F}[V]^G$. This chapter will focus on *modular invariant theory*, the problem of understanding $\mathbf{F}[V]^G$ for V a modular representation. For modular representations, there is no Reynolds operator but the $\mathbf{F}[V]^G$-homomorphism from $\mathbf{F}[V]$ to $\mathbf{F}[V]^G$ induced by the *transfer*, $\mathrm{Tr} := \sum_{g \in G} g$, is still a useful tool.

For more on the invariant theory of finite groups, see Campbell and Wehlau [14], Derksen and Kemper [15], Neusel and Smith [32], and Benson [5].

2. Finite Generation and Noether Normalisation

Suppose that R is a commutative ring with an identity element. R is said to be *Noetherian* if every ascending chain of ideals is stationary. We will make use of the basic properties of Noetherian rings, as outlined in Chapters 6 and 7 of Atiyah and MacDonald [3]. We note that a finitely generated commutative $\mathbf{F}$-algebra is Noetherian and that, if R is Noetherian and M is a finitely generated R-module, then every submodule of M is finitely generated (see Corollary 7.7, Proposition 6.5 and Proposition 6.2 of [3]).

Theorem 1. *If A is a finitely generated commutative $\mathbf{F}$-algebra with an identity and G is a finite group acting on A by algebra automorphisms, then the invariant ring A^G is a finitely generated $\mathbf{F}$-algebra.*

Proof. Suppose A is generated by $\{a_1, a_2, \ldots, a_m\}$. Introduce a new variable t, and, for $i \in \{1, \ldots, m\}$, define
$$F_i(t) := \prod_{g \in G} (t - a_i \cdot g) \in A[t].$$
Observe that $F_i(t) \in A^G[t]$. Expand to get
$$F_i(t) = t^{|G|} + \sum_{j=1}^{|G|} b_{ij} t^{|G|-j}$$
with $b_{ij} \in A^G$. Let B denote the subalgebra of A^G generated by
$$\{b_{ij} \mid i \in \{1, \ldots, m\}, j \in \{1, \ldots |G|\}\}.$$
Since $F_i(a_i) = 0$, we have
$$a_i^{|G|} = -\left(\sum_{j=1}^{|G|} b_{ij} a_i^{|G|-j}\right).$$

Hence, A is generated, as a B-module, by $\{a_1^{e_1} a_2^{e_2} \ldots a_m^{e_m} \mid e_i < |G|\}$. Observe that B is Noetherian, A is a finitely generated B-module, and A^G is a B-submodule of A. Therefore, A^G is a finitely generated B-module. If we take the union of the B-module generators for A^G and the algebra generators for B, we get a finite set of algebra generators for A^G. $\qquad\square$

It follows from the theorem above that $\mathbf{F}[V]^G$ is a finitely generated $\mathbf{F}$-algebra. We can choose the generators to be homogeneous polynomials. The *Noether number*, which we denote by $\beta(V)$, is the largest degree of a polynomial in a minimal homogeneous generating set for $\mathbf{F}[V]^G$. If the representation is non-modular, then $\beta(V) \leq |G|$ (see Fleischmann [19] and Fogarty [23]). In general, as long as $|G| > 1$ and $\dim(V) > 1$,

$$\beta(V) \leq (|G| - 1) \dim(V),$$

(see Symonds [35] or Corollary 2).

Example 3. Suppose the characteristic of $\mathbf{F}$ is a prime p and $G = \langle g \rangle$ is a cyclic group of order p. Let V_2 denote the indecomposable two dimensional $\mathbf{F}G$-module, which is unique up to isomorphism. Choose a basis $\{x, y\}$ for V_2^* so that $x \cdot g = x$ and $y \cdot g = y + x$. Define $N := y^p - x^{p-1}y$. It is not hard to show that $\mathbf{F}[V_2]^G = \mathbf{F}[x, N]$. Let mV_2 denote the direct sum of m copies of V_2 and choose a basis $\{x_i, y_i \mid i \in \{1, \ldots, m\}\}$ for $(mV_2)^*$ so that $x_i \cdot g = x_i$ and $y_i \cdot g = y_i + x_i$. It is easy to verify that $N_i := y_i^p - x_i^{p-1}y_i$ and $u_{ij} := x_i y_j - x_j y_i$ are invariant. Richman [33] conjectured that $\mathbf{F}[mV_2]^G$ is generated by

$$\{x_i, N_i, u_{ij} \mid i, j \in \{1, \ldots, m\}\} \cup \{\mathrm{Tr}(y_1^{e_1} y_2^{e_2} \ldots y_m^{e_m}) \mid 0 \leq e_i \leq p - 1\}$$

and proved that $\mathrm{Tr}\left((y_1 \ldots y_m)^{p-1}\right)$ could not be written as a polynomial in invariants of lower degree for $m > 2$. Campbell and Hughes [12] proved Richman's conjecture. As a consequence, $\beta(mV_2) = m(p - 1)$ for $m > 2$.

Definition 1. Suppose $h_1, \ldots, h_n \in \mathbf{F}[V]^G$ are homogeneous and algebraically independent. Further suppose that $\mathbf{F}[V]$ is a finite module over the subalgebra $\mathcal{P} := \mathbf{F}[h_1, \ldots, h_n]$. Then $\{h_1, \ldots, h_n\}$ is a *homogeneous system of parameters* (or hsop) for $\mathbf{F}[V]^G$ and $\mathcal{P}$ is a *Noether normalisation*.

If $\mathcal{P}$ is a Noether normalisation of $\mathbf{F}[V]^G$, then $\mathcal{P}$ is Noetherian and $\mathbf{F}[V]^G$ is a finite $\mathcal{P}$-module. An often fruitful approach to understanding $\mathbf{F}[V]^G$, is to first find a "nice" Noether normalisation and then to study $\mathbf{F}[V]^G$ as a $\mathcal{P}$-module. For example, suppose V is non-modular, $\mathcal{P} = \mathbf{F}[h_1, \ldots, h_n]$ is a Noether normalisation and $\mathcal{B}$ is a set of $\mathcal{P}$-module generators for $\mathbf{F}[V]$. Then the Reynolds operator is a surjective homomorphism of $\mathcal{P}$-modules and

$$\{\mathcal{R}(a) \mid a \in \mathcal{B}\} \cup \{h_1, \ldots, h_n\}$$

is an $\mathbf{F}$-algebra generating set for $\mathbf{F}[V]^G$.

Example 4. Suppose G acts on V by permuting a basis. Then we can identify G with a subgroup of the symmetric group Σ_n and we have $\mathbf{F}[V]^{\Sigma_n} \subseteq \mathbf{F}[V]^G$. The elementary symmetric polynomial σ_i is the sum of the monomials in the Σ_n-orbit of $x_1 \cdots x_i$ and $\mathbf{F}[V]^{\Sigma_n} = \mathbf{F}[\sigma_1, \ldots, \sigma_n]$ is a Noether normalisation of $\mathbf{F}[V]^G$ (see Exercise 1).

Example 5. If $\mathbf{F}$ is the finite field $\mathbf{F}_q$, then the general linear group $GL_n(\mathbf{F}_q)$ is finite and $\mathbf{F}_q[V]^{GL_n(\mathbf{F}_q)} \subseteq \mathbf{F}_q[V]^G$. The Dickson invariants $d_1, \ldots, d_n$ generate $\mathbf{F}_q[V]^{GL_n(\mathbf{F}_q)}$ and are an hsop for $\mathbf{F}_q[V]^G$ (see Exercise 2).

Theorem 2. *(Dade's Algorithm) Suppose $\mathbf{F}$ is infinite. Then, since V^* is not a finite union of a finite number of proper subspaces, it is possible to choose a basis for V^* so that*

$$x_i \notin \bigcup_{\underline{g} \in G^{i-1}} \mathrm{Span}\{x_1 \cdot g_1, \ldots, x_{i-1} \cdot g_{i-1}\}.$$

With such a choice, $\{x_1 \cdot g_1, \ldots, x_n \cdot g_n\}$ is a basis for V^* for all $\underline{g} \in G^n$. If we define

$$h_i := \prod_{g \in G} x_i \cdot g,$$

then $\{h_1, \ldots, h_n\}$ is an hsop for $\mathbf{F}[V]^G$.

Note that taken together, Theorem 2 and Example 5 show that there exists an hsop for $\mathbf{F}[V]^G$ for every representation V.

Before proving Theorem 2, we introduce methods for identifying hsops. We let $\overline{\mathbf{F}}$ denote the algebraic closure of $\mathbf{F}$ and define $\overline{V} := V \otimes_{\mathbf{F}} \overline{\mathbf{F}}$. Since $\mathbf{F} \subseteq \overline{\mathbf{F}}$, elements of $\mathbf{F}[V]$ represent functions from $\overline{V}$ to $\overline{\mathbf{F}}$. For an ideal $I \subset \mathbf{F}[V]$, define $\mathcal{V}(I) := \{v \in \overline{V} \mid f(v) = 0 \,\forall f \in I\}$. The ideal I is said to be *zero-dimensional* if $\mathcal{V}(I)$ is a finite set.

Proposition 1. *An ideal $I \subseteq \mathbf{F}[V]$ is zero-dimensional if and only if $\mathbf{F}[V]/I$ is a finite dimensional vector space over $\mathbf{F}$. Furthermore, if*

$I = (h_1, \ldots, h_n)$ with the h_i homogeneous, $\mathcal{P} = \mathbf{F}[h_1, \ldots, h_n]$, $\mathbf{F}[V]/I$ is finite dimensional and $b_1, \ldots, b_m \in \mathbf{F}[V]$ are homogeneous polynomials such that $\{b_1 + I, \ldots, b_m + I\}$ is a basis for $\mathbf{F}[V]/I$, then $\{b_1, \ldots, b_m\}$ generates $\mathbf{F}[V]$ as a $\mathcal{P}$-module.

Proof. For a proof that I is zero-dimensional if and only if $\mathbf{F}[V]/I$ is finite dimensional, see Theorem 2.2.7 of Adams and Loustaunau [1]. The fact that $\{b_1, \ldots, b_m\}$ generates $\mathbf{F}[V]$ as a $\mathcal{P}$-module, follows from the Graded Nakayama Lemma, see Proposition 2.10.1 of Campbell and Wehlau [14] or Lemma 3.5.1 of Derksen and Kemper [15].   □

Remark 1. Given homogeneous $h_1, \ldots, h_n \in \mathbf{F}[V]^G$, compute a Gröbner basis for $I = (h_1, \ldots, h_n)\mathbf{F}[V]$ with respect to any monomial order. The set of monomials which are reduced with respect to the Gröbner basis give a generating set for $\mathbf{F}[V]$ as a $\mathcal{P}$-module, where $\mathcal{P}$ is the algebra generated by $\{h_1, \ldots, h_n\}$. If for every x_i there exists an m_i with $x_i^{m_i} \in I$, then the set of reduced monomials is finite and $\{h_1, \ldots, h_n\}$ is an hsop. (For background on Gröbner bases, see Adams and Loustaunau [1].)

Example 6. Suppose $\mathbf{F}$ has characteristic p and G is a p-group. Then $\dim(V^G) \geq 1$ (see Chapter 3 of Alperin [2]) and we can choose a basis $\{x_1, \ldots, x_n\}$ for V^* for which $x_i \cdot g - x_i \in \mathrm{Span}_{\mathbf{F}}\{x_{i+1}, \ldots, x_n\}$ for all $g \in G$ (the matrices representing the elements of G using this basis are upper-triangular and unipotent). Let N_i denote the product of the elements in G-orbit of x_i and let d_i denote the degree of N_i. Choose a monomial order with $x_i > x_{i+1}$. Then the leading term of N_i is $x_i^{d_i}$ and the set $\{N_1, \ldots, N_n\}$ is a Gröbner basis for the ideal $(N_1, \ldots, N_n)\mathbf{F}[V]$. Thus it follows from Remark 1 that $\{N_1, \ldots, N_n\}$ is an hsop and $\mathbf{F}[V]$ is generated, as an $\mathbf{F}[N_1, \ldots, N_n]$-module, by the monomial factors of $x_1^{d_1-1} \cdots x_{n-1}^{d_{n-1}-1}$ (since $N_n = x_n$, we have $d_n = 1$).

Proof of Dade's Algorithm. We will show that $\mathcal{V}(h_1, \ldots, h_n) = \{\underline{0}\}$. The result then follows from Proposition 1. For $v \in V$, we have $h_i(v) = 0$ if and only if $(x_i g_i)(v) = 0$ for some $g_i \in G$. Thus $v \in \mathcal{V}(h_1, \ldots, h_n)$ if and only if there exists $\underline{g} = (g_1, \ldots, g_n) \in G^n$ with $v \in \mathcal{V}(x_1 g_1, x_2 g_2, \ldots, x_n g_n)$. Since $\{g_1 x_1, \ldots, x_n g_n\}$ is a basis for V^*, the system of homogeneous linear equations has a unique solution, giving $\mathcal{V}(x_1 g_1, x_2 g_2, \ldots, x_n g_n) = \{\underline{0}\}$. Therefore $\mathcal{V}(h_1, \ldots, h_n) = \{\underline{0}\}$.   □

We wish to study $\mathbf{F}[V]^G$ as a graded module over a Noether normalisation $\mathcal{P} := \mathbf{F}[h_1, \ldots, h_n]$. It is natural to look at a projective resolution of

$\mathbf{F}[V]^G$. The length of a minimal project resolution is called the *projective dimension*, which we will denote by $\mathrm{pd}_{\mathcal{P}}(\mathbf{F}[V]^G)$.

Proposition 2. *If $\mathcal{P}$ and $\mathcal{P}'$ are two Noether normalisations of $\mathbf{F}[V]^G$, then $\mathrm{pd}_{\mathcal{P}}(\mathbf{F}[V]^G) = \mathrm{pd}_{\mathcal{P}'}(\mathbf{F}[V]^G)$.*

Proposition 2 is essentially a consequence of the Auslander–Buchsbaum formula (see Chapter 19 of Eisenbud [16]) and Lemma 3.7.2 of Derksen and Kemper [15].

We think of projective dimension as a measure of the complexity of a module. Since projective $\mathcal{P}$-modules are necessarily free, if $\mathrm{pd}_{\mathcal{P}}(\mathbf{F}[V]^G) = 0$, then $\mathbf{F}[V]^G$ is a free $\mathcal{P}$-module.

Definition 2. It is usual to define the Cohen–Macaulay (CM) property for a commutative ring using depth and maximal regular sequences, see for example, Bruns and Herzog [11] (or Section 6). However, using the Auslander–Buchsbaum formula and Proposition 2, $\mathbf{F}[V]^G$ is *CM* in the usual sense, if and only if $\mathbf{F}[V]^G$ is a free graded $\mathcal{P}$-module for some (and hence every) Noether normalisation.

Theorem 3. (*Hochster and Eagon [25]*) *If V is a non-modular representation, then $\mathbf{F}[V]^G$ is CM.*

For modular representations, $\mathbf{F}[V]^G$ often fails to be CM.

Theorem 4. (*Kemper [27]*) *If V is a faithful modular representation and mV denotes the direct sum of m copies of V, then for sufficiently large m, the ring $\mathbf{F}[mV]^G$ is not CM. If G is a p-group, then $\mathbf{F}[mV]^G$ is not CM for $m \geq 3$.*

Example 7. Consider the representation mV_2 of Example 3. It follows from Example 6 that $\{x_i, N_i, \mid i = 1, \ldots, m\}$ is an hsop. The relation $x_1 u_{23} - x_2 u_{13} + x_3 u_{12} = 0$ shows that $\mathbf{F}[mV_2]^G$ is not CM for $m \geq 3$.

3. Hilbert Series

For a graded vector space $M = \bigoplus_{i=0}^{\infty} M_i$, the *Hilbert series* of M is the formal power series

$$\mathcal{H}(M,t) := \sum_{i=0}^{\infty} \dim(M_i) t^i.$$

For a polynomial in a single variable, say y, with $\deg(y) = d$,

$$\mathcal{H}(\mathbf{F}[y], t) = 1 + t^d + t^{2d} + \cdots = \frac{1}{1 - t^d}.$$

If M and N are both graded vector spaces, then $\mathcal{H}(M \otimes N, t) = \mathcal{H}(M, t) \cdot \mathcal{H}(N, t)$. Thus

$$\mathcal{H}(\mathbf{F}[V], t) = \frac{1}{(1 - t)^n}$$

and, for a Noether normalisation $\mathcal{P} = \mathbf{F}[h_1, \ldots, h_n]$ with $\deg(h_i) = d_i$,

$$\mathcal{H}(\mathcal{P}, t) = \prod_{i=1}^{n} \frac{1}{1 - t^{d_i}}.$$

It follows from Theorem 3 with $G = \{1\}$ that $\mathbf{F}[V]$ is a free $\mathcal{P}$-module. Applying Proposition 1 gives

$$\mathcal{H}(\mathbf{F}[V]/I, t) = \frac{\mathcal{H}(\mathbf{F}[V], t)}{\mathcal{H}(\mathcal{P}, t)} = \prod_{i=1}^{n} \left(\frac{1 - t^{d_i}}{1 - t} \right) = \prod_{i=1}^{n} (1 + t + \cdots + t^{d_i - 1}),$$

which leads to the following theorem.

Theorem 5. *If $\mathcal{P} = \mathbf{F}[h_1 \ldots, h_n]$ is a Noether normalisation (for some group) then $\mathbf{F}[V]$ is a free graded $\mathcal{P}$-module of rank $\prod_{i=1}^{n} \deg(h_i)$ and top degree $\sum_{i=1}^{n} (\deg(h_i) - 1)$.*

Theorem 6. *(Molien's formula) If $\mathbf{F}$ has characteristic zero, then*

$$\mathcal{H}(\mathbf{F}[V]^G, t) = \frac{1}{|G|} \sum_{g \in G} \frac{1}{\det_V(1 - tg)}.$$

For a proof of Molien's Theorem, see §3.2 of Derksen and Kemper [15] or §2.5 of Benson [5].

Example 8. As in Example 2, consider the usual permutation representation of $A_3 = \langle (1\,2\,3) \rangle$ and assume $\mathbf{F}$ has characteristic zero. Then

$$\rho(A_3) = \left\{ \begin{pmatrix} 1\,0\,0 \\ 0\,1\,0 \\ 0\,0\,1 \end{pmatrix}, \begin{pmatrix} 0\,0\,1 \\ 1\,0\,0 \\ 0\,1\,0 \end{pmatrix}, \begin{pmatrix} 0\,1\,0 \\ 0\,0\,1 \\ 1\,0\,0 \end{pmatrix} \right\}.$$

Thus, $\det_V(1 - 1t) = (1 - t)^3$ and $\det_V(1 - \rho(1\,2\,3)t) = \det_V(1 - \rho(1\,3\,2)t) = (1 - t^3)$. Therefore, using Molien's formula,

$$\mathcal{H}(\mathbf{F}[V]^{A_3}, t) = \frac{1}{3} \left(\frac{1}{(1 - t)^3} + \frac{2}{(1 - t^3)} \right) = \frac{1 + t^3}{(1 - t)(1 - t^2)(1 - t^3)},$$

which is consistent with the observation that $\mathbf{F}[V]^{A_3}$ is a free module over $\mathbf{F}[V]^{\Sigma_3}$ with generators in degrees 0 and 3.

Theorem 7. *Suppose* $\mathcal{P} = \mathbf{F}[h_1, \ldots, h_n]$ *is a Noether normalisation of* $\mathbf{F}[V]^G$ *with* $\deg(h_i) = d_i$. *Then*

$$\mathcal{H}(\mathbf{F}[V]^G, t) = \frac{f(t)}{\prod_{i=1}^{n}(1 - t^{d_i})}$$

for some polynomial $f(t)$ *with integer coefficients. If* $\mathbf{F}[V]^G$ *is CM then the coefficients of* $f(t)$ *are non-negative and* $\mathbf{F}[V]^G$ *is generated, as a* $\mathcal{P}$-*module, by* $r := (\prod_{i=1}^{n} d_i)/|G|$ *homogeneous invariants.*

Proof. A finite free resolution of $\mathbf{F}[V]^G$ as a graded $\mathcal{P}$-module can be used to compute the coefficients of the polynomial $f(t)$. If $\mathbf{F}[V]^G$ is CM, then $f(t)$ is the Hilbert series of $\mathbf{F}[V]^G / ((h_1, \ldots, h_n) \mathbf{F}[V]^G)$ and therefore has non-negative coefficients. The value at $t = 1$ of $(1-t)^n \mathcal{H}(\mathbf{F}[V]^G, t)$ is $1/|G|$, see Theorem 2.4.3 of Benson [5]. Thus, $f(1) = r$ and if $\mathbf{F}[V]^G$ is CM, then $f(1)$ is the number $\mathcal{P}$-module generators. $\square$

4. Integral Extensions and Integral Closure

Suppose S is a subring of a commutative ring R. An element $a \in R$ is *integral* over S if it is a root of a monic polynomial $f(t) \in S[t]$. If every element of R is integral over S, then we say that R is an *integral extension* of S. If R is a finitely generated $\mathbf{F}$-algebra, then R is integral over S if and only if R is a finite S-module. The set of elements in R integral over S is a subring, say $\overline{S}$, known as the *integral closure* of S in R. If $S = \overline{S}$ then S is said to be *integrally closed* in R. A unique factorisation domain (UFD) is integrally closed in its field of fractions. An integral domain which is integrally closed in its field of fractions is called a *normal domain*. See Chapter 5 of Atiyah and MacDonald [3] and §9 of Matsumura [29] for more on integral extensions.

We denote the field of fractions of $\mathbf{F}[V]$ by $\mathbf{F}(V)$. It is clear that the field of fractions of $\mathbf{F}[V]^G$ is contained in $\mathbf{F}(V)^G$. For a finite group, these fields are equal. To see this, write $f/h \in \mathbf{F}(V)^G$ as

$$\frac{f}{h} = \frac{f \prod \{h \cdot g \mid g \in G \setminus \{1\}\}}{\prod \{h \cdot g \mid g \in G\}}$$

and observe that $\prod\{h \cdot g \mid g \in G\} \in \mathbf{F}[V]^G$; thus $f \prod\{h \cdot g \mid g \in G \setminus \{1\}\} \in$ $\mathbf{F}[V]^G$ and every element of $\mathbf{F}(V)^G$ can be written as a ratio of invariant polynomials.

$\mathbf{F}[V]^G$ is integrally closed in its field of fractions (see Proposition 1.1.1 of Benson [5]). This means that if we have a chain of subalgebras $\mathbf{F}[V]^G \subseteq B \subseteq \mathbf{F}(V)^G$ with B a finite $\mathbf{F}[V]^G$-module, then $B = \mathbf{F}[V]^G$.

Theorem 8. *Suppose A is a graded subalgebra of $\mathbf{F}[V]^G$ such that A contains an hsop and a generating set for $\mathbf{F}(V)^G$. If A is integrally closed in its field of fractions, then $A = \mathbf{F}[V]^G$.*

Proof. Since $A \subseteq \mathbf{F}[V]^G$ and A contains a generating set for $\mathbf{F}(V)^G$, the field of fractions of A is $\mathbf{F}(V)^G$. Since A contains an hsop, $\mathbf{F}[V]^G$ is a finite A-module. Therefore, if A integrally closed in its field of fractions, $A = \mathbf{F}[V]^G$. $\square$

Theorem 9. *If V is a faithful representation, $\{h_1, \ldots, h_n\}$ is an hsop for $\mathbf{F}[V]^G$, and $\prod_{i=1}^n \deg(h_i) = |G|$, then $\mathbf{F}[V]^G = \mathbf{F}[h_1, \ldots, h_n]$.*

Proof. From Theorem 5, $\mathbf{F}[V]$ is a free $\mathcal{P}$-module of rank $|G|$ for $\mathcal{P} = \mathbf{F}[h_1, \ldots, h_n]$. This means that the field extension $\mathbf{F}(h_1, \ldots, h_n) \subset \mathbf{F}(V)$ has degree $|G|$. However, the field extension $\mathbf{F}(V)^G \subset \mathbf{F}(V)$ is Galois and therefore also has degree $|G|$. Hence $\mathbf{F}(h_1, \ldots, h_n) = \mathbf{F}(V)^G$. Since a polynomial ring is a UFD, $\mathbf{F}[f_1, \ldots, f_n]$ is integrally closed in its field of fractions and applying Theorem 8 gives $\mathcal{P} = \mathbf{F}[V]^G$. $\square$

Remark 2. The conclusion of Theorem 9 is valid as long as $h_1, \ldots, h_n$ are homogeneous and algebraically independent, see Theorem 3.7.5 of Derksen and Kemper [15].

Example 9. Let V_2 be the indecomposable modular representation of the the cyclic group of order p described in Example 3. Then Theorem 9 and Example 6 combine to show that $\mathbf{F}[V_2]^G = \mathbf{F}[x, N]$. We claim that $\mathbf{F}[2V_2]^G$ is generated by $\mathcal{B} := \{x_1, x_2, N_1, N_2, u_{21}\}$ subject to a single relation: $u_{21}^p = x_2^p N_1 - x_1^p N_2 + (x_1 x_2)^{p-1} u_{21}$. Let A denote the algebra generated by $\mathcal{B}$. From Example 6, $\{x_1, x_2, N_1, N_2\}$ is an hsop. We claim that $\{x_1, x_2, N_1, u_{21}\}$ generates $\mathbf{F}(2V_2)^G$. To see this, observe that $y_2 = (x_2 y_1 - u_{21}) x_1^{-1}$. Thus, $\mathbf{F}[x_1, x_2, y_1, y_2][x_1^{-1}] = \mathbf{F}[x_1, x_2, y_1, u_{21}][x_1^{-1}]$. Furthermore, $\mathbf{F}[x_1, x_2, y_1, u_{21}]^G = \mathbf{F}[x_1, y_1]^G[x_2, u_{21}] = \mathbf{F}[x_1, N_1, x_2, u_{21}]$. Therefore $\mathbf{F}[2V_2]^G[x_1^{-1}] = \mathbf{F}[x_1, x_2, N_1, u_{21}][x_1^{-1}]$ – a much stronger statement than $\mathbf{F}(2V_2)^G = \mathbf{F}(x_1, x_2, N_1, u_{21})$. All that remains is to show that

A is integrally closed in its field of fractions. It is sufficient to show that A is a UFD. Since $A[x_1^{-1}] = \mathbf{F}[x_1, x_2, N_1, u_{12}][x_1^{-1}]$ is a UFD, it is sufficient to show that $x_1 A$ is a prime ideal (see Theorem 20.2 of Matsumura [29]). We leave the details to the reader (or see Theorem 1.1 of Campbell, Shank and Wehlau [13]).

We note that the ring of invariants of a modular representation of a p-group is a UFD (see Theorem 3.8.1 of Campbell and Wehlau [14]).

Example 10. Suppose $\mathbf{F}$ has characteristic not equal to two, $G = \langle g \rangle$ has order two and acts on $V^* = \mathrm{Span}_{\mathbf{F}}\{x, y\}$ by $xg = -x$ and $yg = -y$. Then $\mathbf{F}[V]^G$ is generated by $\{x^2, xy, y^2\}$ subject to the single relation $x^2 y^2 = (xy)^2$. The ring of invariants is clearly not a UFD, an hsop is given by $\{x^2, y^2\}$ and, since $y^2 = (xy)^2/x^2$, the field of fractions is generated by $\{x^2, xy\}$. Furthermore, $\mathbf{F}[V]^G$ is CM — it is the free $\mathbf{F}[x^2, y^2]$-module generated by $\{1, xy\}$.

5. Polynomial Invariants

When is $\mathbf{F}[V]^G$ a polynomial algebra? In other words, for which representations is $\mathbf{F}[V]^G$ generated by an hsop? For non-modular representations, this question is answered by the famous result of Chevalley & Shephard–Todd. For modular representations, the question is open. In this section, we explore the problem for both modular and non-modular representations. We start the section with an outline of the proof of Theorem 11 (the Chevalley–Shepard–Todd–Serre Theorem).

We will call a representation V a *reflection representation* if G is generated by elements $g \in G$ such that $\dim(V^g) = \dim(V) - 1$. (A linear transformation satisfying this property is called a reflection or a pseudo-reflection.) For non-modular representations, $\mathbf{F}[V]^G$ is a polynomial algebra if and only if V is a reflection representation. For modular representations, if $\mathbf{F}[V]^G$ is a polynomial algebra, then V is a reflection representation. However, there are modular reflection representations for which $\mathbf{F}[V]^G$ is not a polynomial algebra, see Example 11.

In order to use powerful categorical techniques from commutative algebra, it is useful to slightly generalise the types of $\mathbf{F}$-algebras we consider. (For background material on categories, functors and homological algebra, see Weibel [37].)

Definition 3. Let $A = \oplus_{i=0}^{\infty} A_i$ be a commutative, finitely generated, graded $\mathbf{F}$-algebra with degree zero component isomorphic to $\mathbf{F}$. Then A

will be called a *graded-connected* **F**-algebra. We define $A_+ := \oplus_{i=1}^{\infty} A_i \subset A$ to be the unique maximal homogeneous ideal of A and with $A - \mathrm{mod}_h$ we will denote the category of finitely generated graded modules $M := \oplus_{i=\ell}^{\infty} M_i$ (with $\ell \in \mathbb{Z}$ and $A_i M_j \subseteq M_{i+j}$) and degree preserving homomorphisms. In other words, $\mathrm{Mor}_{A-\mathrm{mod}_h}(M,N) = \mathrm{Hom}_A(M,N)_0$ for $M, N \in A - \mathrm{mod}_h$. If A is a subalgebra of B, we use $_A B$ to denote B viewed as an A-module.

Recall the definition of free and projective modules in $A - \mathrm{mod}_h$:

$X \in A-\mathrm{mod}_h$ is *free* $\iff$ $X \cong \oplus_{i=1}^{n} A e_i$ with homogeneous generators of degree d_i and $A e_i \cong A[-d_i]$, the shifted graded module.

$P \in A - \mathrm{mod}_h$ is *projective* $\iff$ every commutative diagram of the form:

with exact row can be completed by a morphism in $A - \mathrm{mod}_h$.

It is well known (and easy to see) that P is projective if and only if P is a direct summand of a free module.

There are two useful functors:

$$A - \mathrm{mod}_h \xrightarrow{\ E\ } \mathbf{F} - \mathrm{mod}_h \xrightarrow{\ F\ } A - \mathrm{mod}_h$$

with $E(M) := \mathbf{F} \otimes_A M \cong M/A_+ M$ (a finite dimensional graded vector space) and $F(N) := A \otimes_{\mathbf{F}} N \cong \oplus_{i=1}^{k} A \nu_i$ (a graded free A-module), where $N = \oplus_{i=1}^{k} \mathbf{F} \nu_i$ is a free graded vector space. Since these are tensor functors they are right exact with the obvious effect on morphisms. Note that $EF(N) \cong N$ in a canonical way and there is a (non-canonical) surjective map $FE(M) = A \otimes_{\mathbf{F}} M/A_+ M \to M \to 0$: indeed, if $\{m_1, \ldots, m_\mu\}$ is a (minimal) generating set for $M \in A-\mathrm{mod}_h$, then $M/A_+ M = \oplus_{i=1}^{\mu} \mathbf{F}\bar{m}_i$ and $FE(M) \cong \oplus_{i=1}^{\mu} A e_i$ with $\deg(e_i) = \deg(m_i)$, maps onto M via $a \cdot e_i \mapsto a \cdot m_i$.

The following graded version of Nakayama's Lemma is well known.

Lemma 1. *(Nakayama's Lemma) For $M \in A - \mathrm{mod}_h$,*

$$M = A_+ M \iff M = 0.$$

This implies that the functor E is faithful in the sense that $E(M) = 0$ if and only if $M = 0$. The following is a distinctive property of the category

$A - \text{mod}_h$, which it shares with the category of modules over commutative local rings.

Lemma 2. *Let* $P \in A - \text{mod}_h$, *then* P *is projective* $\iff P$ *is free.*

Proof. For projective P the diagram

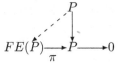

splits, i.e., $FE(P) \cong P \oplus \ker(\pi)$. Now apply E to obtain $E(P) \cong EFE(P) \cong E(P) \oplus E(\ker(\pi))$. It follows $E(\ker(\pi)) = 0 \Rightarrow \ker(\pi) = 0 \Rightarrow P \cong FE(P)$, hence free.   □

We will need a few general facts from commutative algebra. It follows from a theorem of Serre that a local Noetherian ring R has finite global dimension if and only if R is a regular ring (see Matsumura [29] Theorem 19.2) This implies that for a graded-connected **F**-algebra,

- A is a polynomial algebra $\iff A$ has finite global dimension.

Indeed, a polynomial algebra has finite global dimension by Hilbert's Syzygy theorem. On the other hand, if A has finite global dimension, so does the localisation A_{A_+}, which then is a regular local ring. The fact that A is a polynomial algebra follows from Exercise 19.1 of Matsumura [29].

Lemma 3. *Let* S *be a graded-connected polynomial algebra and* A *a graded-connected subalgebra of* S *such that* S *is a finite* A-*module. Then* A *is a polynomial algebra if and only if* S *is a free* A-*module.*

Proof. Suppose A is a polynomial algebra. Then A is generated by an hsop and, since S is CM, $_AS$ is free.

Suppose $_AS$ is free. Let $P^* \to A/A_+ \to 0$ be a minimal free resolution. Then $P^* \otimes_A S \to A/A_+ \otimes_A S \cong S/A_+S \to 0$ is a minimal free resolution in $S - \text{mod}_h$, using the fact that $_AS$ free and therefore faithfully flat. Hence $P^N \otimes_A S = 0$ for $N >> 0$, so $P^N = 0$, which implies that A has finite global dimension and, therefore, A is a polynomial algebra.   □

Lemma 4. *(Freeness criterion) Suppose* $M \in A - \text{mod}_h$ *and the canonical multiplication map*

$$A_+ \otimes_A M \xrightarrow{\mu} M, \ a \otimes m \mapsto am$$

is injective. Then M *is free.*

Proof. The short exact sequence

$$0 \longrightarrow K \overset{j}{\longrightarrow} FE(M) \overset{p}{\longrightarrow} M \longrightarrow 0,$$

gives rise to the exact sequence

$$A_+ \otimes K \longrightarrow A_+ \otimes FE(M) \longrightarrow A_+ \otimes M \longrightarrow 0$$

and the commutative diagram: with exact rows:

$$
\begin{array}{ccccccccc}
0 & \longrightarrow & K & \overset{j}{\longrightarrow} & FE(M) & \overset{p}{\longrightarrow} & M & \longrightarrow & 0 \\
& & \uparrow a & & \downarrow b & & \uparrow c & & \\
& & A_+ \otimes K & \longrightarrow & A_+ \otimes FE(M) & \longrightarrow & A_+ \otimes M & \longrightarrow & 0.
\end{array}
$$

Note that b is injective because $FE(M)$ is free and c is injective by hypothesis. We need to show that a is surjective, then $K = A_+ K$, so $K = 0$ (by Nakayama's Lemma). Since E is right exact we obtain the exact sequence:

$$EK \overset{\bar{j}}{\longrightarrow} EFE(M) \overset{\cong}{\longrightarrow} EM \longrightarrow 0.$$

The proof can be finished by a diagram chase. $\qquad\square$

For the rest of this section, S will be a graded-connected **F**-algebra and G will be a finite group, acting faithfully on S by graded algebra automorphisms. Let $\text{Spec}_h(S)$ denote the set of homogeneous prime ideals in S, then one defines for every $\mathfrak{P} \in \text{Spec}_h(S)$ the *inertia group* of $\mathfrak{P}$ as $I_G(\mathfrak{P}) := \{g \in G \mid S(g-1) \subseteq \mathfrak{P}\}$. This is a normal subgroup of the *decomposition group* $D_G(\mathfrak{P}) := \{g \in G \mid \mathfrak{P}g = \mathfrak{P}\}$ as defined in Bourbaki [6] (p. 330). Note that the group $D_G(\mathfrak{P})$ acts on the residue class field $k(\mathfrak{P}) := \text{Quot}(S/\mathfrak{P})$ with kernel $I_G(\mathfrak{P})$. Let $\text{Spec}_{1,h}(S)$ denote the subset of homogeneous prime ideals of height one in S. We call an element g a (generalised) *reflection on S* if $g \in I_G(\mathfrak{P})$ for some $\mathfrak{P} \in \text{Spec}_{1,h}(S)$. Moreover we define

$$W := W_G := \langle I_G(\mathfrak{P}) \mid \mathfrak{P} \in \text{Spec}_{1,h}(\mathbf{F}[V]) \rangle \trianglelefteq G.$$

Since G permutes the set $\text{Spec}_{1,h}(S)$, it is clear that W_G is a normal subgroup of G, called the *(generalised) reflection subgroup of G on S*. The reason why $g \in I_G(\mathfrak{P})$ is called a "generalised reflection on S" is the following:

Assume S is also a UFD, then every $\mathfrak{P} \in \mathrm{Spec}_{1,h}(S)$ is principal, so $\mathfrak{P} = (f)$ for some homogeneous irreducible element $f \in S$ (see Matsumura [29] Theorem 20.1). In particular, if $S = \mathbf{F}[V]$ and $G \leq \mathrm{GL}(V)$, then

$$g \in I_G(\mathfrak{P}) \iff (V^*)(g-1) \subseteq (f)$$

and $I_G(\mathfrak{P}) = 1$ if $\deg(f) > 1$. If $\deg(f) = 1$, then $f \in V^* \setminus \{\underline{0}\}$, and $H := \ker(f|_V)$ is a hyperplane in V. For $v \in H$, for any $\phi \in V^*$, we have $\phi((g-1)v) = (\phi(g-1))(v) = cf(v) = 0$ (for some $c \in \mathbf{F}$). Thus $(g-1)v = 0$ and, hence, $H \subseteq V^g$. Since the action is faithful $H = V^g$ and g acts as a reflection on V. So if $S = \mathbf{F}[V]$ with $G \leq \mathrm{GL}(V)$, then W_G is the normal subgroup of G generated by the linear reflections in G.

Theorem 10. *Let S be a graded-connected $\mathbf{F}$-algebra which is also a UFD and let G be a finite group acting faithfully on S by graded $\mathbf{F}$-algebra automorphisms. Assume that $G = W_G$ and $|G| \in \mathbf{F}^*$. Then $_SG S$ is free.*

Proof. (See also Bourbaki [7] Theorem 1, p. 110.) Set $A := S^G$. For $M \in A - \mathrm{mod}_h$, we define

$$T(M) := \ker(A_+ \otimes_A M \xrightarrow{\mu} M).$$

The idea is to show that $T(S) = 0$ and to use Lemma 4. For $\alpha \in \mathrm{Hom}_A(M, N)_i$ we have $(\mathrm{id} \otimes \alpha)(T(M)) \subseteq T(N)$, hence we can define

$$T(\alpha) = (\mathrm{id} \otimes \alpha)|_{T(M)} \in \mathrm{Hom}_A(T(M), T(N))_i.$$

Clearly, $T(\alpha \circ \beta) = T(\alpha) \circ T(\beta)$, so T is a degree preserving additive functor from $A - \mathrm{mod}_h$ to itself. If $g \in I_G(\mathfrak{P})$ with $\mathfrak{P} = (f) \in \mathrm{Spec}_{1,h}(S)$ and $f \in S_+$ with $\deg(f) = i > 0$, then define $\theta_g \in \mathrm{Hom}_A(S, S)_{-i}$ by

$$(s)(g-1) = (s)\theta_g \cdot f$$

for $s \in S$. Hence, $T(g) - T(\mathrm{id}) = T(\theta_g) \circ T(\mu_f)$ where μ_f is the homomorphism "multiply by f" and $\deg(T(\theta_g)) < 0$. If $T(S) \neq 0$, then the set of G-fixed points $T(S)^G \neq 0$. Indeed: pick $0 \neq t \in T(S)$ of minimal degree; for every $g \in I_G(\mathfrak{P})$ and $\mathfrak{P} \in \mathrm{Spec}_{1,h}(S)$, we have $(t)(T(g) - T(\mathrm{id})) = (t)T(\theta_g) \circ T(\mu_f)$, with $t' = (t)T(\theta_g) \in T(S)$ of degree $< \deg(t)$. Hence $t' = 0$ and $t \in T(S)^g$. Since $G = W_G$, we conclude that $t \in T(S)^G$. It remains to show that $T(S)^G = 0$: Let $t = \sum_i r_i^+ \otimes a_i \in (A_+ \otimes_A S)^G \cap T(S)$, then $t = 1/|G| \sum_{g \in G}(t)(\mathrm{id} \otimes g) = 1/|G| \sum_i r_i^+ \otimes \mathrm{tr}(a_i) = 1/|G| \sum_i r_i^+ \mathrm{tr}(a_i) \otimes 1 = \mu(t) \otimes 1 = 0 \in A_+ \otimes_A A$. This shows that $T(S) = 0$, so the multiplication map $A_+ \otimes_A S \to S$ is injective and $_A S$ is free by Lemma 4. $\qquad\square$

The main result of this section is the following theorem.

Theorem 11. *Let S be a graded-connected polynomial algebra on which the finite group G acts by graded algebra automorphisms (e.g., $S = \mathbf{F}[V]$ and $G \leq \mathrm{GL}(V)$, but the case where S is generated in higher degrees is included). Consider the following statements:*

(1) $_{S^G}S$ *is free.*
(2) S^G *is a polynomial algebra.*
(3) G *is generated by reflections on S (i.e., $G = W_G$).*

Then the following implications hold: (1) $\Longleftrightarrow$ (2) $\Rightarrow$ (3).
If $|G| \in \mathbf{F}^*$ then (1) $\Longleftrightarrow$ (2) $\Longleftrightarrow$ (3).

Proof. The equivalence "(1) $\Longleftrightarrow$ (2)" has already been shown in Lemma 3 and "(3) $\Rightarrow$ (1)", when $|G| \in \mathbf{F}^*$, follows from Theorem 10. It remains to show "(2) $\Rightarrow$ (3)" ("Serre's theorem"). This needs several ideas and from algebraic number theory, which we are collecting in the results below. Using those the proof finishes as follows: Assume (2) and set $W := W_G$. Then by Propositions 4 and 3, we see that $_{S^G}S^W$ is free and "unramified in height one". Proposition 3 then implies $S^W = S^G$, hence $\mathrm{Quot}(S)^W = \mathrm{Quot}(S)^G$ and $W = G$ follows by standard Galois theory.   □

Let $A \leq S$ be a finite extension of graded-connected $\mathbf{F}$-algebras which are also normal domains. We also assume that the extension of the quotient fields $\mathbb{K} := \mathrm{Quot}(A) \leq \mathrm{Quot}(S) =: \mathbb{L}$ is (finite) separable. In the applications later, S will have a graded G-action with $A = S^G$, so the extension of quotient fields will be Galois with group G. A *fractional ideal I of S* is a finitely generated S-submodule of $\mathbb{L}$ and for each fractional ideal I, one defines $I^{-1} := \{\ell \in \mathbb{L} \mid \ell I \subseteq S\}$, which is again a fractional ideal. The fractional ideal I is called *divisorial*, if it is the intersection of *principal fractional ideals* $S \cdot q$ with $q \in \mathbb{L}$. Divisorial fractional ideals are determined by their localisations at homogeneous height one prime ideals, i.e., if I, J are (homogeneous) divisorial fractional ideals, then $I = J$ if and only if $I_{\mathfrak{P}} = J_{\mathfrak{P}} \trianglelefteq S_{\mathfrak{P}}$ for all (homogeneous) prime ideals $\mathfrak{P} \trianglelefteq S$ of height one. Note that, since S is a normal domain, $S_{\mathfrak{P}}$ is a discrete valuation ring, so

$$I = J \iff \nu_{\mathfrak{P}}(I) = \nu_{\mathfrak{P}}(J) \ \forall \mathfrak{P} \in \mathrm{Spec}_{1,h}(S).$$

If S is a UFD, then every (homogeneous) divisorial fractional ideal I is principal, i.e., $I = S \cdot q_0$ for some homogeneous element $q_0 \in \mathbb{L}$. It turns out that there is a divisorial ideal, the *Dedekind different*, $\mathcal{D}_{S,A} \trianglelefteq S$, satisfying

$$\mathcal{D}_{S,A}^{-1} = \{\ell \in \mathbb{L} \mid \mathrm{Tr}(\ell B) \in A\}.$$

Here Tr denotes the usual trace operator defined for the separable field extension $\mathbb{L} \geq \mathbb{K}$. If S has G-action with $A = S^G$, e.g., $S = \mathbf{F}[V]$ and $A = \mathbf{F}[V]^G$, then Tr coincides with the group theoretic trace operator

$$\mathrm{Tr} : \mathbb{L} \to \mathbb{K} = \mathbb{L}^G, \; \ell \mapsto \sum_{g \in G} \ell^g.$$

The Dedekind different plays a central role in *ramification theory* which studies the local structure of the ring extension $A \leq S$. Let $\mathfrak{P} \in \mathrm{Spec}_h(S)$ with $\mathfrak{p} := \mathfrak{P} \cap A$, then $\mathfrak{P}$ is called *unramified over A*, if $\mathfrak{p}S_{\mathfrak{P}} = \mathfrak{P}S_{\mathfrak{P}}$ and the extension of residue class fields $S_{\mathfrak{P}}/\mathfrak{P}S_{\mathfrak{P}} \geq A_{\mathfrak{p}}/\mathfrak{p}A_{\mathfrak{p}}$ is separable. The ring extension $A \leq S$ is called *unramified* (in height one) if every $\mathfrak{P} \in \mathrm{Spec}_h(S)$ ($\mathfrak{P} \in \mathrm{Spec}_{h,1}(S)$) is unramified. It is well known that $A \leq S$ is unramified in height one if and only if $\mathcal{D}_{S,A} = S$ (see Proposition 3.6 of Auslander and Buchsbaum [4]). If moreover $_AS$ is projective, then $A \leq S$ is unramified $\iff \mathcal{D}_{S,A} = S$ (see Corollary 3.7 of [4]).

Let $\mu \colon S \otimes_A S \to S$ be the multiplication map with kernel $J \trianglelefteq S \otimes_A S$. Since μ is surjective, the annihilator $\mathrm{ann}_{S \otimes_A S}(J) \trianglelefteq S \otimes_A S$ is mapped onto a homogeneous ideal $\mathcal{D}_{S,A,hom} := \mu(\mathrm{ann}_{S \otimes_A S}(J)) \trianglelefteq S$, which is called the *homological different* of the extension $_AS$. If $_AS$ is finitely generated and projective, then $\mathcal{D}_{S,A,hom} = \mathcal{D}_{S,A}$ (Proposition 3.3 of [4]).

Proposition 3. *Assume $_AS$ is free and unramified, then $A = S$.*

Proof. By the assumption, $1_S \in \mathcal{D}_{S,A} = \mathcal{D}_{S,A,hom} = \mu(\mathrm{ann}_{S \otimes_A S}(J))$, so $1_{S \otimes_A S} - x \in \mathrm{ann}_{S \otimes_A S}(J)$ for some $x \in \ker(\mu) = J$ and for every $j \in J$ we get $j = xj$, hence $J = J^2$. Since A and S are graded-connected, $J_0 = 0$, hence $J = J_+ := J \cap (S \otimes_A S)_+$ and therefore $J = J^2$ implies $J = 0$. It follows that $S \otimes_A S \cong S$. Since $_AS$ is free of finite rank, this implies that $_AS$ is free of rank one, hence $A = S$. $\square$

Now let $A \leq S \leq T$ be finite extensions of graded-connected $\mathbf{F}$-algebras, which are also normal domains. Then we have the following generalised version of Dedekind's tower theorem from algebraic number theory (see Benson [5] Lemma 3.10.1, p. 39):

Lemma 5. $\mathcal{D}_{T,A} = \overline{\mathcal{D}_{T,S} \cdot \mathcal{D}_{S,A}}.$

Here $\bar{J}$ denotes the "divisorial closure" of $J \trianglelefteq T$, i.e., the smallest divisorial ideal of T containing J.

Lemma 6. *Suppose S is a normal domain which is also a graded-connected $\mathbf{F}$-algebra. Further suppose that a finite group G acts faithfully on S by graded algebra automorphisms. Then $\mathcal{D}_{S,S^G} = \mathcal{D}_{S,S^{W_G}}$ and $\mathcal{D}_{S^{W_G},S^G} = S^{W_G}$.*

Proof. Set $A := S^G$, $W := W_G$ and let $\mathfrak{P} \in \mathrm{Spec}_{1,h}(S)$, then by standard ramification theory the extension $A_{\mathfrak{P} \cap A} \leq S_{\mathfrak{p}}^{I_G(\mathfrak{P})}$ (with $\mathfrak{p} := \mathfrak{P} \cap S^{I_G(\mathfrak{P})}$) is finite and unramified, hence $\mathcal{D}_{S_{\mathfrak{p}}^{I_G(\mathfrak{P})}, A_{\mathfrak{P} \cap A}} = S_{\mathfrak{p}}^{I_G(\mathfrak{P})}$. It follows that

$$\nu_{\mathfrak{P}}(\mathcal{D}_{S,A}) = \nu_{\mathfrak{P}}\left(\mathcal{D}_{S_{\mathfrak{P}},S_{\mathfrak{p}}^{I_G(\mathfrak{P})}} \cdot \mathcal{D}_{S_{\mathfrak{p}}^{I_G(\mathfrak{P})}, A_{\mathfrak{P} \cap A}}\right)$$
$$= \nu_{\mathfrak{P}}(\mathcal{D}_{S_{\mathfrak{P}},S_{\mathfrak{p}}^{I_G(\mathfrak{P})}})$$
$$= \nu_{\mathfrak{P}}(\mathfrak{D}_{S,S^{I_G(\mathfrak{P})}}).$$

Since $I_G(\mathfrak{P}) \leq W$, we obtain

$$\nu_{\mathfrak{P}}(\mathcal{D}_{S,A}) = \nu_{\mathfrak{P}}(\mathfrak{D}_{S,S^{I_G(\mathfrak{P})}}) = \nu_{\mathfrak{P}}(\mathfrak{D}_{S,S^{I_W(\mathfrak{P})}}) = \nu_{\mathfrak{P}}(\mathcal{D}_{S,S^W}).$$

Hence, $\mathcal{D}_{S,A} = \mathcal{D}_{S,S^W}$. From this and Lemma 5, we obtain $\nu_{\mathfrak{q}}(\mathcal{D}_{S^W,A}) = 0$ for all $\mathfrak{q} \in \mathrm{Spec}_{1,h}(S^W)$, hence $\mathcal{D}_{S^W,A} = S^W$. $\qquad\square$

Let $C \leq B \leq A$ be commutative unitary subrings. Then ${}_C A \cong {}_C B \otimes_B A$, hence by the adjointness of tensor and Hom functors there is an isomorphism Ψ of Abelian groups:

$$\mathrm{Hom}_C({}_C A, C) = \mathrm{Hom}_C({}_C B \otimes_B A, C) \cong \mathrm{Hom}_B({}_B A, \mathrm{Hom}_C({}_C B, C));$$

$$\Psi: \ \phi \mapsto (a \mapsto (b \mapsto \phi(b \cdot a))); \quad \Psi^{-1}: \ \gamma \mapsto (a \mapsto \gamma(a)(1)).$$

Both Abelian groups are also A–C-bimodules and Ψ is an isomorphism of bimodules. Indeed:

$$\Psi(a'\phi c)\,(a)(b) = a'\phi c(ba) = \phi(baa') \cdot c = (\Psi(\phi)(aa')(b)) \cdot c$$
$$= ((\Psi(\phi)(aa'))c)(b) = (a'\Psi(\phi)c)(a)(b).$$

Lemma 7. *Let $\lambda_{(A,B)} \in \mathrm{Hom}_B({}_B A, B)$ and $\lambda_{(B,C)} \in \mathrm{Hom}_C({}_C B, C)$ be such that the maps*

$$\phi^{(A,B)}: \ A \longrightarrow \mathrm{Hom}_B({}_B A, B), \ a \mapsto a \cdot \lambda_{(A,B)} \text{ and}$$

$$\phi^{(B,C)}: \ B \to \mathrm{Hom}_C({}_C B, C), \ b \mapsto b \cdot \lambda_{(B,C)},$$

are isomorphisms of $A - B$ and $B - C$-bimodules. Then the map

$$\phi^{(A,C)} : A \longrightarrow \mathrm{Hom}_C({}_C A, C), \ a \mapsto a \cdot \lambda_{(A,C)},$$

with $\lambda_{(A,C)} = \lambda_{(B,C)} \circ \lambda_{(A,B)}$ is an isomorphism of $A - C$-bimodules.

Proof. Consider the bijection

$$(\phi^{(B,C)})_* \circ \phi^{(A,B)} : A \to \mathrm{Hom}_B(A, \mathrm{Hom}_C(B, C)),$$

then for $a, a' \in A$ and $b \in B$ we have

$$(\phi^{(B,C)})_* \circ \phi^{(A,B)}(a)(a')(b) = \lambda_{(B,C)}(\lambda_{(A,B)}(aa')b).$$

Hence $(\phi^{(B,C)})_* \circ \phi^{(A,B)}(a) = \lambda_{(B,C)}(\lambda_{(A,B)}(a \cdot) \cdot)$. Now define

$$\phi^{(A,C)} := \Psi^{-1} \circ (\phi^{(B,C)})_* \circ \phi^{(A,B)} : A \mapsto \mathrm{Hom}_C({}_C A, C).$$

Then $\phi^{(A,C)}$ is bijective and $\phi^{(A,C)}(a)(a') = \Psi^{-1}((\phi^{(B,C)})_* \circ \phi^{(A,B)}(a))$ $(a') = \lambda_{(B,C)}(\lambda_{(A,B)}(a \cdot a') \cdot 1) = (a \cdot \lambda_{(B,C)} \circ \lambda_{(A,B)})(a')$. So for $\lambda_{(A,C)} := \lambda_{(B,C)} \circ \lambda_{(A,B)} \in \mathrm{Hom}_C(A, C)$ the map $\phi^{(A,C)}$ is an isomorphism of A–C-bimodules. □

Proposition 4. *Let S be a graded-connected $\mathbf{F}$-algebra which is also a UFD. Assume moreover that S^G is a direct summand of the module ${}_{S^G} S$. Then $p = \mathrm{char}(\mathbf{F})$ does not divide $|G/W_G|$, S^G is a direct summand of ${}_{S^G} S^{W_G}$ and ${}_{S^G} S^{W_G}$ is a direct summand of ${}_{S^G} S$. In particular, if ${}_{S^G} S$ is free, then so is ${}_{S^G} S^{W_G}$.*

Proof. As before, we set $A := S^G$ and $W := W_G$. It is clear that S^G is a direct summand of ${}_{S^G} S^W$. Since S is a UFD, $\mathcal{D}_{S,S^W}$ is a principal ideal, hence $\mathcal{D}_{S,S^W}^{-1}$ is a principal fractional ideal and therefore $\mathrm{Hom}_{S^W}(S, S^W) \cong S \cdot \lambda'$ is a cyclic S-module generated by a homogeneous element λ'. From Lemma 6, we see that $\mathrm{Hom}_A(S^W, A) \cong S^W \cdot \mathrm{Tr}_{\bar{G}}$ with (relative) trace $\mathrm{Tr}_{\bar{G}} : S^W \to S^G$ for $\bar{G} := G/W$. It follows from Lemma 7 that $\mathrm{Hom}_A(S, A) \cong S \cdot \lambda$ with $\lambda = \mathrm{Tr}_{\bar{G}} \circ \lambda'$. Since A is a direct summand of ${}_A S$, there is $s_1 \in S$ such that $s_1 \cdot \lambda$ is the projection $S \to A$, hence $1 = \lambda(s_1 s) = \mathrm{Tr}_{\bar{G}}(\lambda'(s_1 s))$ for some $s \in S$. It follows that $\deg(\lambda'(s_1 s)) = 0$, so $\lambda'(s_1 s) \in \mathbf{F}$ is a non-zero constant c with $\mathrm{Tr}_{\bar{G}}(c) = |\bar{G}| \cdot c \neq 0$. We can assume $\lambda'(s_1 s) = 1$, so $\lambda' : S \to S^W$ is a surjective morphism of S^W-modules, which therefore splits, showing that S^W is a direct summand of ${}_{S^W} S$. Finally, if ${}_A S$ is free, A is a summand of ${}_A S$, hence ${}_A S^W$ is also a summand of ${}_A S$ and therefore projective. It now follows from Lemma 2 that ${}_A S^W$ is free. □

The following example, taken from Campbell and Wehlau [14] demonstrates that in the modular situation the implication "(3) $\Rightarrow$ (2)" in Theorem 11 is false, in other words, there are groups generated by pseudo reflections whose ring of invariants is not a polynomial ring:

Example 11. Let $\mathbf{F}$ be a field of characteristic p and define

$$G = \left\{ \begin{pmatrix} 1 & 0 & a & c \\ 0 & 1 & c & b \\ 0 & 0 & 1 & 0 \\ 0 & 0 & 0 & 1 \end{pmatrix} \mid a, b, c \in \mathbf{F}_p \right\}.$$

Then G is generated by reflections but $\mathbf{F}[V]^G$ is not a polynomial ring (see Theorem 8.2.4 of Campbell and Wehlau [14]).

Remark 3. Kemper and Malle [28] classified all finite irreducible modular groups $G \leq \mathrm{GL}(V)$ with polynomial rings of invariants. It turns out that for irreducible G, $\mathbf{F}[V]^G$ is polynomial if and only if $G = W_G$ and for every $v \in \overline{V} := \overline{\mathbf{F}} \otimes_{\mathbf{F}} V$, the stabiliser G_v has a polynomial ring of invariants $\mathbf{F}[V]^{G_v}$.

Assume that $G \leq \mathrm{GL}(V)$ is a p-group with $p = \mathrm{char}(\mathbf{F})$, then the reflections in G are *transvections* of order p. Hence if G is not generated by elements of order p, $\mathbf{F}[V]^G$ cannot be a polynomial ring.

Example 12. Consider $S := \mathbf{F}_2[x_1, x_2, x_3]$ and $G = \langle g \rangle$ with $x_1 g = x_1$, $x_2 g = x_2 + x_1$ and $x_3 g = x_3 + x_2$. Then G is cyclic of order four and S^G is generated by $\{x_1, f_2, f_3, f_4\}$ with $f_2 := x_1 x_2 + x_2^2$, $f_3 := x_1^2 x_3 + x_1 x_2^2 + x_1 x_3^2 + x_2^3$ and $f_4 := x_1^2 x_2 x_3 + x_1^2 x_3^2 + x_1 x_2^2 x_3 + x_1 x_2 x_3^2 + x_2^2 x_3^2 + x_3^4$ subject to the relation $x_1^2 f_4 - f_2^3 - x_1 f_2 f_3 - f_3^2 = 0$. It follows that $f_4 = \frac{1}{x_1^2}(f_2^3 + x_1 f_2 f_3 + f_3^2)$, hence the "localised" ring of invariants is a "localised polynomial ring":

$$S^G[x_1^{-1}] = \mathbb{F}_2[x_1^{\pm 1}, f_2, f_3].$$

Set $x := x_1$ and let D_x be the "dehomogenisation" $D_x := (S[1/x])_0 \cong S/(x-1)S$, then D_x is a polynomial ring with faithful non-homogeneous G-action, such that the ring of invariants D_x^G is isomorphic to the polynomial ring $\mathbf{F}[f_2/x^{\deg(f_2)}, f_3/x^{\deg(f_3)}]$. This shows that the implication "(2) $\Rightarrow$ (3)" in Theorem 11 requires the group action to be graded and is false for "non-linear" group actions on polynomial rings. It has been shown by Fleischmann and Woodcock [22] that every finite p-group has a faithful (mildly nonlinear) action on a polynomial ring with polynomial ring of invariants.

5.1. *p-groups*

In this subsection, $P \leq \mathrm{GL}(V)$ will denote a p-group with $p = \mathrm{char}(\mathbf{F})$. It is well known that the P-fixed point space V^P is non-zero if $V \neq 0$. Repeatedly using this property we can always find dual bases $\mathcal{B} = \{v_1, \ldots, v_n\}$ of V and $\mathcal{B}^* = \{x_1, \ldots, x_n\}$ of V^* such that the matrices $M_{\mathcal{B}}(g)$ are upper triangular unipotent for every $g \in P$ (compare with Example 6). For every $x \in V^*$ set

$$P_x := \mathrm{Stab}_P(x) = \{g \in G \mid xg = x\}, \quad N(x) := N_{P_x}^P(x) = \prod_{y \in xP} y,$$

the "orbit product" of x, and $P_i := \cap_{j \neq i} P_{x_j}$. Then $P_i \leq \mathrm{GL}(V)$ is a "one-row-subgroup" of P:

$$\begin{pmatrix} 1 & 0 & 0 & \cdots & & 0 & 0 & 0 & 0 \\ 0 & 1 & 0 & \cdots & & 0 & 0 & 0 & 0 \\ & & & \vdots & & & & & \\ 0 & 0 & \cdots & 0 & 1 & 0 & 0 & 0 & \cdots & 0 \\ 0 & 0 & \cdots & 0 & 0 & 1 & * & * & \cdots & * \\ 0 & 0 & \cdots & 0 & 0 & 0 & 1 & 0 & \cdots & 0 \\ & & & & \vdots & & & & & \\ 0 & 0 & 0 & \cdots & & & 0 & 0 & 1 & 0 \\ 0 & 0 & 0 & \cdots & & & 0 & 0 & 0 & 1 \end{pmatrix}.$$

The group P is called a *Nakajima* group with respect to $\mathcal{B}$, if $P = P_n P_{n-1} \cdots P_1$.

Theorem 12. *P is Nakajima group with respect to the basis $\mathcal{B}$ if and only if*

$$\mathbf{F}[V]^P = \mathbf{F}[N(x_1), \ldots, N(x_n)].$$

Proof. We know from Example 6 that the $N(x_i)$'s form an hsop. Moreover $\deg(N(x_i)) = |P_i|$ with $\prod_i |P_i| = |P|$. Now the claim follows from Theorem 9. The converse was proven by Yinglin Wu (see Theorem 8.0.11 of Campbell and Wehlau [14].). $\qquad \square$

Remark 4. In 1980, H. Nakajima proved that if $\mathbf{F} = \mathbf{F}_p$ the converse is also true, i.e., if $G \leq \mathrm{GL}(V)$ is defined over $\mathbf{F}_p$ and $\mathbf{F}[V]^G$ is a polynomial ring, then G is a Nakajima group (Nakajima [30 and 31]). This converse is false for $\mathbf{F} = \mathbf{F}_{p^n}$, $n > 1$ as shown by a counter-example due to Stong (see Campbell and Wehlau [14] p. 147).

The image of the transfer is an ideal in the ring of invariants. Motivated by by a range of examples, which included all Nakajima groups, Shank and Wehlau [34] made the following conjecture.

Conjecture 1. *If V is a modular representation of a p-group, P, then $\mathbf{F}[V]^P$ is a polynomial algebra if and only if the image of the transfer is a principal ideal.*

Suppose the image of the transfer is generated by $a = \mathrm{Tr}(h)$ (for homogeneous $h \in \mathbf{F}[V]$). Then define $\rho : \mathbf{F}[V] \to \mathbf{F}[V]^G$ by $\rho(f) = \mathrm{Tr}(h \cdot f)/a$. Then ρ is a degree preserving surjective $\mathbf{F}[V]^G$-module map and $\mathbf{F}[V] = \mathbf{F}[V]^G \oplus \ker(\rho)$ is a decomposition of $\mathbf{F}[V]^G$-modules. If $\mathbf{F}[V]^G$ is a polynomial algebra, then $\mathbf{F}[V]$ is a free $\mathbf{F}[V]^G$-module and it is easy to see that $\mathbf{F}[V]^G$ is a direct summand of $\mathbf{F}[V]$. Broer [10] showed that, for a modular representation of a p-group, $\mathbf{F}[V]^P$ is a direct summand if and only if the image of the transfer is principal and he extended Conjecture 1 to the following.

Conjecture 2. *For a reflection representation, $\mathbf{F}[V]^G$ is a polynomial algebra if and only if $\mathbf{F}[V]^G$ is a direct summand of $\mathbf{F}[V]$.*

6. On the Depth of Modular Rings of Invariants

We have seen that, for a modular representation, the ring of invariants is not in general CM. If $\mathbf{F}[V]^G$ is not CM, then the projective dimension of $\mathbf{F}[V]^G$ as a module over a Noether normalisation gives a measure of how far $\mathbf{F}[V]^G$ is from being CM. Using the Auslander–Buchsbaum formula, the projective dimension is the Krull dimension minus the depth. Thus computing the depth is a way to measure how far the ring is from being CM. Recall that the *grade* of an ideal J on a module M, denoted by $\mathrm{grade}(J, M)$, is the length of a maximal M-regular sequence in J. The *depth* of $\mathbf{F}[V]^G$ is $\mathrm{grade}(\mathbf{F}[V]_+^G, \mathbf{F}[V]^G)$ (see Bruns and Herzog [11] for details). Throughout this section, we assume $\mathrm{char}(\mathbf{F}) = p > 0$. We continue to work with graded-connected $\mathbf{F}$-algebras, see Definition 3. Note that by our definition, a graded-connected $\mathbf{F}$-algebra is Noetherian.

Lemma 8. *Let $B \leq A$ be an extension of graded-connected $\mathbf{F}$-algebras and let $J \lhd A$ be a proper homogeneous ideal such that $J^k \subseteq (J \cap B)A$ for some $k \in \mathbb{N}$. Let $N \in A - \mathrm{mod}_h$, then*

$$\mathrm{grade}(J, N) = \mathrm{grade}((J \cap B)A, N) = \mathrm{grade}(J \cap B, N).$$

If moreover $_BN \cong M \oplus K$ with homogeneous B-submodules M and K (with $M \neq 0$), then

$$\text{grade}(J, N) \leq \text{grade}(J \cap B, M).$$

In particular, if $_BA$ is finitely generated and B is direct summand of $_BA$, then $\text{depth}(A) \leq \text{depth}(B)$.

Proof. Note that $J \leq A_+$, so by Nakayama's Lemma we can assume that $JN < N$. The first equality follows from $J^k \subseteq (J \cap B)A$ and the well known fact that if $\underline{j} := (j_1, \ldots, j_d)$ is a regular sequence in J on N, then so is $(j_1^{e_1}, \ldots, j_d^{e_d})$ for any exponents $e_i \in \mathbb{N}$. Let $(j_1, \ldots, j_d)$ be a maximal N-regular sequence in $J \cap B$. Then $J \cap B$ is contained in some associated prime $\mathfrak{q} \in Ass_B(N/\underline{j}N)$, hence $(J \cap B)A$ annihilates a non-zero element in $N/\underline{j}N$, so $\underline{j}$ is also maximally N-regular in $(J \cap B)A$. Since $(J \cap B)N = (J \cap B)AN < N$, all maximal N-regular sequences in $J \cap B$ as well as those in $(J \cap B)A$ have the same length, which proves the second equality. Now assume that $_BN \cong M \oplus K$. Let $m \in M$ be such that $j_i m \in (j_1, \ldots, j_{i-1})M$ for $i \leq d$. Then $m \in (j_1, \ldots, j_{i-1})N \cap M = (j_1, \ldots, j_{i-1})M$, showing that $(j_1, \ldots, j_d)$ is M-regular. □

Consider a subgroup $H \leq G$ with $p \nmid [G : H]$ and assume that G acts on the algebra A, then the *relative transfer*

$$\text{Tr}_H^G : A^H \to A^G, \ a \mapsto \sum_{g \in G/H} (a)g$$

is a split epimorphism of A^G modules, so A^G is a direct summand of $_{A^G}A^H$. Using Lemma 8 we obtain the following generalisation of Theorem 3.

Corollary 1. *Suppose $H \leq G$ and p does not divide $[G : H]$. Then $\text{depth } \mathbf{F}[V]^H \leq \text{depth } \mathbf{F}[V]^G$. If P is a Sylow p-subgroup of G and $\mathbf{F}[V]^P$ is CM, so is $\mathbf{F}[V]^G$.*

Example 13. Consider the usual permutation action of $G = \Sigma_p$ on $V = \mathbf{F}^p$. Then $\mathbf{F}[V]^G$ is polynomial and hence CM. Thus $\text{depth}(\mathbf{F}[V]^G) = \dim(V) = p$. However, with $P = \langle (1\,2 \ldots p) \rangle$ and $p \geq 5$, $\text{depth}(\mathbf{F}[V]^P) = 3$ by a result of Ellingsrud and Skjelbred [17] (or see Theorem 13 below).

A group G is called *p-nilpotent* if it has a normal p-complement, i.e., a normal subgroup N of order coprime to p, such that G/N is a p-group (which then is isomorphic to a Sylow p-subgroup of G).

Theorem 13. *(Fleischmann, Kemper and Shank [20]) If G is p-nilpotent with cyclic Sylow p-subgroup $P \leq G$, then*

$$\mathrm{depth}(\mathbf{F}[V]^G) = \min\left\{\dim(V^P) + 2,\, \dim(V)\right\},$$

where V^P denotes the P-fixed point space.

For $P = G$ cyclic, this result was originally obtained by Ellingsrud and Skjelbred [17]. They also show that for any group G with Sylow p-subgroup P,

$$\mathrm{depth}\, \mathbf{F}[V]^G \geq \min\left\{\dim(V^P) + 2, \dim(V)\right\}. \tag{1}$$

Hence, $\dim(V^P)$ is a lower bound for the depth of $\mathbf{F}[V]^G$. Let $\mathfrak{J} \trianglelefteq \mathbf{F}[V]$ denote the ideal vanishing on the fixed point space V^P. Then

$$\mathfrak{J} = (\mathbf{F}[V](g-1) \mid g \in P)S = \sqrt{\mathbf{F}[V]_{<P}^G}, \tag{2}$$

where $\mathbf{F}[V]_{<P}^G := \sum_{Q<P} \mathrm{Tr}_Q^G(\mathbf{F}[V]^Q) \trianglelefteq \mathbf{F}[V]^G$ (see Fleischmann [18] or Fleischmann, Kemper and Shank [20]). The following result shows that the "missing part" of the depth in the inequality (1) is provided by a regular sequence in the ideal generated by relative transfer invariants induced from p-subgroups.

Theorem 14. *(Fleischmann and Shank [21]) Let P be a Sylow p-subgroup of $G \leq \mathrm{GL}(V)$. Then*

$$\mathrm{depth}\, \mathbf{F}[V]^G = \mathrm{grade}(\mathbf{F}[V]_{<P}^G, \mathbf{F}[V]^G) + \dim(V^P).$$

We close this section with the statement of a celebrated theorem that proved a conjecture by Landweber and Stong about the special role of Dickson invariants in modular invariant theory.

Theorem 15. *(Bourguiba and Zarati [8]; also see Henn [24]) For $A = \mathbf{F}_q[V]$ with V an $\mathbf{F}_qG$ -module and $G \leq \mathrm{GL}(V)$, one can use the Dickson invariants (see Exercise 2) as a test sequence to measure the depth of A^G. More precisely, the depth of A^G is the largest ℓ such that the sequence $d_1, \ldots, d_\ell$ is A^G-regular.*

7. Castelnuovo Regularity and Finite Decomposition Type

The polynomial ring $\mathbf{F}[V]$ is an infinite dimensional $\mathbf{F}G$-module, decomposed in a natural way by its homogeneous components. These homogeneous components have been studied extensively and are a useful way of constructing new representations from old (see for example, pp. 14–16 of Alperin [2]). In modular representation theory it is well known that, although there are only finitely many isomorphism types of simple $\mathbf{F}G$-modules, there are, for most G, infinitely many isomorphism types of *indecomposable* ones. It is therefore natural to ask if only a finite number of indecomposable $\mathbf{F}G$-modules appear as direct summands of the homogeneous components of $\mathbf{F}[V]$. Again, throughout this section, we assume $\mathrm{char}(\mathbf{F}) = p > 0$.

Definition 4. The $\mathbf{F}G$-module M is said to be of *finite decomposition type*, if M is a sum of finite dimensional indecomposable $\mathbf{F}G$-modules with only finitely many different isomorphism types appearing in the decomposition.

Karagueuzian and Symonds [26] proved that $\mathbf{F}[V]$ is of finite decomposition type for any finite $G \leq \mathrm{GL}(V)$. This was achieved using explicit but very complicated calculations. The result was then applied by Symonds to analyse the Castelnuovo regularity of $\mathbf{F}[V]$ which led to a proof of a long conjectured degree bound for the Noether number for modular representations (see Corollary 2). In a recent paper, Symonds gave a new very conceptual proof of all these results, applying links between standard cohomology, local cohomology and the cohomology of Čech-complexes. In this section, we will give a brief sketch of his proof, concentrating on its "invariant theory" part, while restricting to basic definitions and facts about the background cohomology theories. Proofs for those facts are easily found in the referred literature.

Let A denote a graded-connected $\mathbf{F}$-algebra and let $A-Mod_h$ denote the category of $\mathbb{Z}$-graded A-modules. For $M \in A - Mod_h$, we define $\mathrm{beg}(M) := \inf\{i \in \mathbb{Z} \mid M_i \neq 0\} \in \mathbb{Z} \cup \{\pm\infty\}$ and $\mathrm{end}(M) := \sup\{i \in \mathbb{Z} \mid M_i \neq 0\} \in \mathbb{Z} \cup \{\pm\infty\}$. So if for example $A := \mathbf{F}[X]$ is a univariate polynomial ring over the field $\mathbf{F}$, then $\mathrm{beg}(_A A) = 0$ and $\mathrm{end}(_A A) = \infty$, $\mathrm{beg}(_A \mathbf{F}[X^z]) = z$ for $z \in \mathbb{Z}$ and $\mathrm{end}(_A A/(X^n)) = n - 1$ for $n \in \mathbb{N}$.

For a fixed homogeneous ideal $I \triangleleft A$, one can define the *local cohomology* modules as

$$H_I^q(M) := \varinjlim \mathrm{Ext}_{A,h}^q(A/I^i, M).$$

Here, $\mathrm{Ext}^q_{A,h}(A/I^i, M)$ denotes the graded Ext groups as defined, for example, in §1.5 of Bruns and Herzog [11]. Since A/I is a finitely generated module, $\mathrm{Ext}^q_A(A/I^i, M)$ is the same as $\mathrm{Ext}^q_{A,h}(A/I^i, M)$ with grading forgotten. It follows that $H^q_I(M)$ inherits a natural grading from M and A/I. These modules give rise to the corresponding additive *local cohomology functor* H^*_I (with respect to the ideal $I \trianglelefteq A$), which is the right derived functor of

$$M \mapsto H^0_I(M) = \{m \in M \mid I^n m = 0 \text{ for some } n \in \mathbb{N}\}.$$

For more details on (graded) local cohomology theory, we refer to Brodmann and Sharp [9] or Bruns and Herzog [11].

The connection between local cohomology and the graded Ext groups is given by Grothendieck's local duality theorem. The following is a special formulation in the case where A is a polynomial ring (this is sufficient for our purposes):

Theorem 16. *Let* $\mathcal{P} := \mathbf{F}[f_1, f_2, \ldots, f_n]$ *with* $d_\ell := \deg(f_\ell) > 0$ *be a polynomial ring. Then for every* $M \in \mathcal{P} - \mathrm{mod}_h$ *and* $i \in \mathbb{Z}$ *there is an isomorphism of graded* $\mathbf{F}$*-vector spaces*

$$\mathrm{Hom}_\mathbf{F}(H^i_{\mathcal{P}_+}(M), \mathbf{F}) \cong \mathrm{Ext}^{n-i}_{\mathcal{P},h}(M, \mathcal{P})[a],$$

with $a := -\sum_{\ell=1}^n d_\ell$. *Moreover for every* $j \in \mathbb{Z}$, *the local cohomology group* $H^i_{\mathcal{P}_+}(M)_j$ *is a finite dimensional* $\mathbf{F}$*-vector space satisfying the identity:*

$$\dim_\mathbf{F}(H^i_{\mathcal{P}_+}(M)_j) = \dim_\mathbf{F}(\mathrm{Ext}^{n-i}_{\mathcal{P},h}(M, \mathcal{P})_{a-j}).$$

Proof. See §13.4.6 of Brodmann and Sharp [9], where all $d_\ell = 1$. We leave the straightforward generalisation as a useful exercise for the reader. □

Definition 5. We set $\mathfrak{m} := A_+$ and define the following numerical invariants of $M \in A - \mathrm{mod}_h$:

- $\epsilon_i(M) := \mathrm{beg}(\mathrm{Ext}^i_{A,h}(M, A))$;
- $\alpha_i(M) := \mathrm{end}(H^i_\mathfrak{m}(M))$;
- $\mathrm{Reg}(M) := \sup\{-\epsilon_i(M) - i \mid i \geq 0\}$;
- $\mathrm{Reg}_{LC}(M) := \sup\{\alpha_i(M) + i \mid i \geq 0\}$;
- $\mathrm{hreg}(M) := \sup\{-\epsilon_i(M) \mid i \geq 0\}$;
- $\mathrm{hreg}_{LC}(M) := \sup\{\alpha_i(M) \mid i \geq 0\}$.

The invariant $\mathrm{Reg}(M)$ is called the *Castelnuovo–Mumford regularity* of M (see §15.2.9 of Brodmann and Sharp [9]).

Remark 5. It is well known that $H_{\mathfrak{m}}^i(M) = 0$ for all $i > \mathrm{Dim}(A)$, (where $\mathrm{Dim}(A)$ denotes the Krull dimension of A) hence it follows from the definition that $\mathrm{hreg}_{LC}(M) + \mathrm{Dim}(A) \geq \mathrm{Reg}_{LC}(M)$.

We extend the definition of the *Noether number* to $M \in A - \mathrm{mod}_h$. Let $\mathfrak{M} = \{\mu_1, \ldots, \mu_m\} \subseteq M$ be a minimal set of homogeneous generators such that $\deg(\mu_1) \leq \cdots \leq \deg(\mu_m) =: \gamma$. Assume that $\mathfrak{N} = \{\nu_1, \ldots, \nu_\ell\} \subseteq M$ is an arbitrary set of homogeneous generators of M; order $\mathfrak{N}$ in such a way that $\deg(\nu_1) \leq \cdots \leq \deg(\nu_\ell)$, then $\gamma \leq \deg(\nu_\ell)$. Indeed, assume $\deg(\nu_\ell) < \gamma$, then

$$\mathfrak{N} \subseteq \langle \mu_i \mid \deg(\mu_i) \leq \deg(\nu_\ell) < \gamma \rangle_A \leq \langle \mu_1, \ldots, \mu_{m-1} \rangle_A,$$

in contradiction to the minimality of $\mathfrak{M}$. We therefore can define the Noether number $\beta(M) := \gamma$, which is independent of the particular minimal set of homogeneous generators and is indeed an invariant of $M \in A - \mathrm{mod}_h$.

Definition 6. For $M \in A - \mathrm{mod}_h$ we define

$$\mathrm{Rreg}(M) := \sup\{\beta(F_i) - i \mid i \in \mathbb{N}_0\},$$

where $F_* \to M \to 0$ is a minimal projective resolution of M in $A - \mathrm{mod}_h$.

Theorem 17. *Let $\mathcal{P} := \mathbf{F}[f_1, \ldots, f_n]$ and $M \in \mathcal{P} - \mathrm{mod}_h$ be as in Theorem 16. Then*

$$\mathrm{Reg}(M) = \mathrm{Rreg}(M) = \mathrm{Reg}_{LC}(M) - a - n.$$

Proof. For the first equation, see Proposition 20.16 of Eisenbud [16], where again all d_ℓ are 1. As before, we leave the straightforward generalisation as an exercise to the reader. The second equation follows from Theorem 16. $\qquad\square$

Suppose G acts on A by graded automorphisms and consider the class $A - G - \mathrm{mod}_h$ of finitely generated graded A-modules, which are also $\mathbf{F}G$-modules in such a way that for all $g \in G$, $a \in A$ and $m \in M$: $g(am) = g(a) \cdot gm$ (where $g(a) := (a)g^{-1}$, if the initial action of G on A is a right action). In analogue to Theorem 14 we define $A_{<G}^G := \sum_{Q<G} \mathrm{Tr}_Q^G(A^Q) \trianglelefteq A^G$.

Using a close connection between local cohomology and the cohomology of Čech-complexes, which we will not define here, Symonds proved the following fundamental theorem.

Theorem 18. *(Symonds [36]) Let $M \in A - G - \mathrm{mod}_h$ and let $t \in \mathbb{N}$ be an integer such that $\mathrm{hreg}_{LC}(M) \leq t$. Suppose that for each p-subgroup $P \leq G$*

there exist homogeneous elements $\mathbf{y} := (y_1, \ldots, y_r)$ and $\mathbf{z} := (z_1, \ldots, z_s)$ of A_+^P such that the following hypotheses hold:

(a) $y_i \in \sqrt{A_{<P}^P}$ for $i = 1, \ldots, r$;

(b) $\sqrt{(\mathbf{y}, \mathbf{z})A} = A_+$;

(c) there is a $\mathbf{F}[\mathbf{y}] - P$-module T and a $\mathbf{F}[\mathbf{z}] - P$-module U such that $M_{>t} \cong (T \otimes_{\mathbf{F}} U)_{>t}$ as $\mathbf{F}[\mathbf{y}, \mathbf{z}] - P$-modules and the action of P on U is trivial.

Then:

(a) $\mathrm{hreg}_{LC}(M^G) \le t$;

(b) M is of finite decomposition type.

Proof. See Theorem 4.1 of Symonds [36]. For background on the connection between local cohomology and Čech-cohomology, see also Brodmann and Sharp [9], or Bruns and Herzog [11].   □

Theorem 19. *The $\mathbf{F}G$-module $\mathbf{F}[V]$ is of finite decomposition type with*

$$\mathrm{hreg}_{LC}(\mathbf{F}[V]^G) \le -n \text{ and } \mathrm{Reg}_{LC}(\mathbf{F}[V]^G) \le 0,$$

where $n = \dim_{\mathbf{F}}(V)$.

Proof. Set $S := \mathbf{F}[V]$; by Remark 5, we have $\mathrm{Reg}_{LC}(S^G) \le \mathrm{hreg}_{LC}(S^G) + n$, so $\mathrm{hreg}_{LC}(S^G) \le -n$ implies $\mathrm{Reg}_{LC}(S^G) \le 0$. Therefore it suffices to verify the hypotheses of Theorem 18 with $t = -n$. Let $P \le G$ be a p-group and $\mathcal{B} = (v_1, \ldots, v_n)$ an ordered basis of V such that the (left) P action on V is given by lower triangular matrices and $(v_{r+1}, \ldots, v_n)$ is a basis of the P-fixed point space V^P with $\dim_{\mathbf{F}}(V^P) = s$ and $r = n - s$.

In other words, $\mathcal{M}_{\mathcal{B}}(g) = \begin{pmatrix} 1 & & & & & \\ \star & 1 & & & & \\ \star & \star & 1 & & & \\ \star & \star & \star & 1 & & \\ \star & \star & \star & \mathbf{0} & 1 & \\ \star & \star & \star & \mathbf{0} & \mathbf{0} & 1 \end{pmatrix}$ with non-zero-entries only in the first r positions of each row. The right action of g on V^* is described by right action of $\mathcal{M}_{\mathcal{B}}(g)$ on the standard rows. In particular we see that $(x_i)g = x_i + \sum_{j<i, j \le r} \mathcal{M}_{\mathcal{B}}(g)_{i,j} x_j$ for all $i = 1, \ldots, n$. Now define for $i = 1, \ldots, r$, $y_i := N_{P/P_{x_i}}(x_i) := \prod_{g \in P:P_{x_i}} (x_i)g$, the orbit product of x_i and for $j = 1, \ldots, s$, $z_j := N_{P/P_{x_{r+j}}}(x_{r+j})$. Since $\langle x_1, \ldots, x_r \rangle = (V^P)^\perp \le V^*$, it

follows from equation (2) that the y_i lie in $\sqrt{S_{<P}^P}$. Let $a_j := \deg(z_j)$ for $j = 1, \ldots, s$. We now define

$$T := \langle \underline{x}^\alpha \mid \alpha \in \mathbb{N}_0^n, \; \alpha_{r+j} < a_j, \; \forall 1 \leq j \leq s \rangle_{\mathbf{F}},$$

the vector space spanned by all monomials with x_{r+j}-degree strictly less that a_j. It follows from the above description of the P-action, that T is an infinite dimensional $\mathbf{F}P$-module. It also follows that every y_i only involves variables x_ℓ with $\ell \leq i$, hence T is an $\mathbf{F}[\mathbf{y}] - P$-module. We also define $U := \mathbf{F}[\mathbf{z}]$ with trivial P-action. As a vector space, $T \cong S/(x_{r+1}^{a_1}, \ldots, x_{r+s}^{a_s})$, so its Hilbert-series satisfies $H_S(t) = \frac{H_T(t)}{\prod_{j=1}^{s}(1-t^{a_j})}$. Hence $H_{T \otimes_{\mathbf{F}} U}(t) = H_T(t) \cdot H_U(t) = H_S(t)$. It therefore suffices to show that the standard map $T \otimes_{\mathbf{F}} U \to S$ given by the embeddings $T \hookrightarrow S$ and $U \hookrightarrow S$ and multiplication is surjective. We choose a graded term-order on S with $x_1 < x_2 < \cdots < x_n$. Then $z_j = x_{r+j}^{a_j} + X_j$, with every monomial in X_j being strictly smaller than $x_{r+j}^{a_j}$. Now let a be the smallest monomial that is not in the image of the map above. If each x_{r+j} for $1 \leq j \leq s$ appears in a to a power strictly less than a_j, then $a \in T$ and therefore in the image. Otherwise, let j be the largest value with $1 \leq j \leq s$ such that $x_{r+j}^{a_j}$ divides a; then $a = x_{r+j}^{a_j} \cdot b = (z_j - X_j)b$. Since $b < a$, it is in the image, hence so is $z_j b$. Every monomial appearing in $X_j b$ is a product of a monomial of X_j and one of b, hence strictly smaller than $x_{r+j}^{a_j} \cdot b = a$ and therefore in the image. It follows that $X_j b$ is in the image and therefore a is in the image, a contradiction. It follows that $T \otimes_{\mathbf{F}} U \cong S$. Thus all hypothesis of Theorem 18 are verified for $M = S$ and for any $t \geq -n$. $\qquad\square$

Obviously, $\beta(M) \leq \mathrm{Rreg}(M)$, which is the clue to the following.

Corollary 2. *Suppose V is a modular representation of G of dimension n and that d_i are the degrees of the elements of an hsop for $\mathbf{F}[V]^G$. Let $\mathcal{P}$ denote the corresponding Noether normalisation. Then $\beta({}_{\mathcal{P}}\mathbf{F}[V]^G) \leq \sum_{i=1}^{n}(d_i - 1)$ and*

$$\beta(\mathbf{F}[V]^G) \leq \max\{|G|, \dim_{\mathbf{F}}(V) \cdot (|G| - 1)\}.$$

Proof. Clearly, $\beta({}_{\mathcal{P}}\mathbf{F}[V]^G) \leq \mathrm{Rreg}({}_{\mathcal{P}}\mathbf{F}[V]^G)$. Since $\mathrm{Reg}_{LC}(\mathbf{F}[V]^G) \leq 0$, the first claim follows from Theorem 17. It is well known that $\beta(\mathbf{F}[V]^G)$ does not change under field extensions, so we can assume $\mathbf{F}$ to be algebraically closed. In that case there is an hsop such that all $d_i \leq |G|$ by Dade's algorithm (see Theorem 2). $\qquad\square$

8. Exercises

Exercise 1. Let the symmetric group Σ_n act on $\mathbf{F}[V] = \mathbf{F}[x_1, x_2, \ldots, x_n]$ by permuting the variables. Define σ_i to be the sum of the elements in the Σ_n-orbit of $x_1 x_2 \cdots x_i$. Show that $\mathbf{F}[V]^{\Sigma_n} = \mathbf{F}[\sigma_1, \ldots, \sigma_n]$.

Solution: We could prove this directly by showing that every symmetric polynomial can be written as a polynomial in the σ_i. To do this, use a lexicographic monomial order and the fact that the orbit sums of monomials give a vector space basis for $\mathbf{F}[V]^{\Sigma_n}$ to show that for any $f \in \mathbf{F}[V]^G$, the lead monomial of f can be written as a product of the lead monomials of the σ_i. Use this to reduce f and the result follows by minimal counter-example.

Another approach would be to use Theorem 9. Since $\prod_{i=1}^{n} \deg(\sigma_i) = n! = |\Sigma_n|$, it is sufficient to show that $\{\sigma_1, \ldots, \sigma_n\}$ is an hsop. Using Proposition 1, it is sufficient to show that $\mathcal{V}(\sigma_1, \ldots, \sigma_n) = \{\underline{0}\}$. Since $\sigma_n = x_1 x_2 \cdots x_n$, for any $v \in \mathcal{V}(\sigma_1, \ldots, \sigma_n)$, we have $x_i(v) = 0$ for some i. Without loss of generality, we can assume $i = n$. The result follows by induction (the $n = 1$ case is clear). Alternatively, consider the polynomial $F(t) := \prod_{i=1}^{n}(t - x_i) \in \mathbf{F}[V][t]$. The coefficients of F are, up to sign, the σ_i and the roots of F are the x_i. Thus $x_i^n \in (\sigma_1, \ldots, \sigma_n)\mathbf{F}[V]$ and $\{\sigma_1, \ldots, \sigma_n\}$ is an hsop by Remark 1 (or use this as another way to show $\mathcal{V}(\sigma_1, \ldots, \sigma_n) = \{\underline{0}\}$).

Exercise 2. Let V denote a vector space of dimension n over $\mathbf{F}_q$, the field with $q = p^r$ elements. Define $F_V(t) := \prod_{\phi \in V^*}(t - \phi)$ and

$$
\Delta_n(t) := \det \begin{pmatrix}
x_1 & x_2 & \cdots & x_n & t \\
x_1^q & x_2^q & \cdots & x_n^q & t^q \\
 & & \vdots & & \\
x_1^{q^n} & x_2^{q^n} & \cdots & x_n^{q^n} & t^{q^n}
\end{pmatrix}.
$$

(a) How are $F_V(t)$ and $\Delta_n(t)$ related?
   Hint: compare roots and leading terms.
(b) Show that $F_V(t) = \sum_{i=0}^{n} d_{n-i} t^{q^i}$ for some $d_j \in \mathbf{F}[V]$.
   (The d_i are known as the Dickson invariants.)
(c) Show that $\mathbf{F}_q[V]^{GL(V)} = \mathbf{F}_q[d_1, \ldots, d_n]$.
(d) Let $U_q(n)$ denote the subgroup of upper-triangular unipotent matrices in $GL_n(\mathbf{F}_q)$. Define $N(x_i) := \prod \{x \cdot g \mid g \in U_q(n)\}$
   (the product of the elements in the orbit of x_i).
   Explain why $\mathbf{F}_q[x_1, \ldots, x_n]^{U_q(n)} = \mathbf{F}_q[N(x_1), \ldots, N(x_n)]$.

Show that $N(x_i) = \sum_{j=0}^{i-1} c_{i-j-1} x_i^{q^j}$ for some $c_k \in \mathbf{F}[x_1, \ldots, x_n]$.
What can you say about the c_k?

Solution: (a) We consider both $F_V(t)$ and $\Delta_n(t)$ as polynomials in t with coefficients in $\mathbf{F}_q[V] = \mathbf{F}_q[x_1, \ldots, x_n]$. $F_V(t)$ has degree q^n (as a polynomials in t) and the roots of $F_V(t)$ are the q^n elements of V^*. Since taking a q-power is $\mathbf{F}_q$-linear, the elements of V^* are also roots of $\Delta_n(t)$. Note that the coefficient of t^{q^n} in $\Delta_n(t)$ is $\Delta_{n-1}(x_n)$. Thus, as long as $\Delta_{n-1}(x_n) \neq 0$, we have $F_V(t) = \Delta_n(t)/\Delta_{n-1}(x_n)$. We can prove that $\Delta_{n-1}(x_n) \neq 0$ by induction starting with $\Delta_1(x_2) = x_1 x_2^q - x_2 x_1^q$.

(b) Since $F_V(t) = \Delta_n(t)/\Delta_{n-1}(x_n)$, only the q^i powers of t appear in the expansion of $F_V(t)$.

(c) Since $GL(V)$ permutes the roots of $F_V(t)$, we have $F_V(t) \in \mathbf{F}[V]^{GL(V)}[t]$ and $d_k \in \mathbf{F}[V]^{GL(V)}$. By construction $\deg(d_k) = q^n - q^{n-k}$. Therefore $\prod_{j=1}^{n} \deg(d_j) = \prod_{i=0}^{n-1} (q^n - q^i) = |GL(V)|$. Therefore, to apply Theorem 9, we need only show that $\{d_1, \ldots, d_n\}$ is an hsop. Since x_i is a root of $F_V(t)$, we have $x_i^{q^n} \in (d_1, \ldots, d_n)\mathbf{F}[V]$ and the result follows from Remark 1.

(d) It follows from Example 6 that $\{N(x_1), \ldots, N(x_n)\}$ is an hsop. Observe that the orbit of x_i is $x_i + \text{Span}_{\mathbf{F}_q}\{x_{i+1}, \ldots, x_n\}$. Therefore $\deg(N(x_i)) = q^{n-i}$ and $\prod_{i=1}^{n} \deg(N(x_i)) = \prod_{j=1}^{n-1} q^j = |U_q(n)|$. Thus, it follows from Theorem 9, that $\mathbf{F}_q[x_1, \ldots, x_n]^{U_q(n)} = \mathbf{F}_q[N(x_1), \ldots, N(x_n)]$. Observe that $N(x_i) = F_{W_i}(x_i)$ where $W_i^* = \text{Span}_{\mathbf{F}_q}\{x_{i-1}, \ldots, x_n\}$. Thus only the q^j powers of x_i appear in the expansion of $N(x_i)$. The c_k appearing in the expansion of $N(x_i)$ are the Dickson invariants for the vector space W_i.

Exercise 3. Let R be a normal integral domain and $G \leq \text{Aut}(R)$. Show that R^G is also a normal integral domain.

Solution: Let $x = f/g \in \text{Quot}(R^G) \leq \text{Quot}(R)$ be integral over R^G. Then x is integral over R, hence $x \in R \cap \text{Quot}(R^G) = R^G$.

Exercise 4. Let $G \leq GL(V)$ be a finite group, $A := \mathbf{F}[V]^G \leq S := \mathbf{F}[V]$ and $\bar{S} := S/A_+ S$. Show:

(a) $\dim_{\mathbf{F}}(\bar{S}) = |G| \iff {}_A S$ is free $\iff A$ is a polynomial algebra.
(b) Assume ${}_A S$ is free. Then $\bar{S}$ is isomorphic to the regular module ${}_{\mathbf{F}G}\mathbf{F}G$ if and only if $|G| \cdot 1_{\mathbf{F}} \neq 0$.

(Hints: For (a) use the functors E and F above, with respect to A. For (b) use normal basis theorem from Galois theory, Maschke's theorem from representation theory and use the fact that for finite dimensional $X, Y \in \mathbf{F}G - \text{mod}$ and any field extension $\mathbb{L} \geq \mathbf{F}$, we have $\mathbb{L} \otimes_{\mathbf{F}} X \cong \mathbb{L} \otimes_{\mathbf{F}} Y$ as $\mathbb{L}G$-modules $\iff X \cong Y$ as $\mathbf{F}G$-modules.)

Solution: (a) Assume $\dim_{\mathbf{F}}(\bar{S}) = |G|$. Then

$$F(\bar{S}) = A \otimes_{\mathbf{F}} S/A_+S \xrightarrow{\pi} S$$

is a (minimal) free cover of $_AS$ with rank $|G|$ (as an A-module, not necessarily as an AG-module). Let $\mathbb{L} := \text{Quot}(S)$ with $\mathbb{K} := \text{Quot}(A) = \mathbb{L}^G$, then, by exactness of localisation the map $\mathbb{L} \otimes_A \pi : \mathbb{K} \otimes_A F(\bar{S}) \to \mathbb{L}$ is a surjective homomorphism of $\mathbb{K}$-vector spaces of dimension $|G|$, hence an isomorphism. It follows that π is injective and therefore an isomorphism, too.

Assume that $_AS$ is free. Then $_AS = \oplus_{i=1}^{|G|} Ae_i$ with $[\mathbb{L} : \mathbb{K}] = |G|$, since $\mathbb{L} \geq \mathbb{K}$ is Galois' with group G. It follows that

$$\bar{S} := E(_AS) \cong \oplus_{i=1}^{|G|} Ae_i/A^+e_i \cong \oplus_{i=1}^{|G|} \mathbf{F}e_i \in \mathbf{F}G - \text{mod}.$$

(b) Note that $\mathbf{F} \cong \bar{S}_0$ is a direct summand of $\bar{S}$. If $\bar{S} \cong {}_{\mathbf{F}G}\mathbf{F}G$, then the trivial module $\mathbf{F}G$-module $\mathbf{F}$ is projective, which implies $|G| \cdot 1_{\mathbf{F}} \neq 0$.

Now assume that $|G| \in \mathbf{F}^*$, then $S = A_+S \oplus X$ with $X \cong \bar{S} \in \mathbf{F}G - \text{mod}$. By Galois theory, there is a normal-basis generator $\ell \in \mathbb{L}$, hence for $Y := \oplus_{g \in G} \mathbf{F}\, {}^g\ell \subseteq \mathbb{L}$ we have $Y \cong {}_{\mathbf{F}G}\mathbf{F}G$ and

$$\mathbb{K} \otimes_{\mathbf{F}} Y \cong \mathbb{L} \cong \mathbb{K} \otimes_{\mathbf{F}} X \in \mathbb{K}G - \text{mod}.$$

It follows that $Y \cong X \cong \bar{S}$.

References

[1] W. W. Adams and P. Loustaunau, *An Introduction to Gröbner Bases*, American Mathematical Society, 1994.

[2] J. L. Alperin, *Local Representation Theory*, Cambridge University Press, 1986.

[3] M. F. Atiyah and I. G. MacDonald, *Introduction to Commutative Algebra*, Addison-Wesley, 1969.

[4] M. Auslander and D. A. Buchsbaum, On ramification theory in Noetherian rings, *Amer. J. of Math* **81** (1959), 749–765.

[5] D. J. Benson, *Polynomial Invariants of Finite Groups*, Cambridge University Press, 1993.

[6] N. Bourbaki, *Commutative Algebra, Chapters 1–7*, Springer-Verlag, 1989.

[7] N. Bourbaki, *Lie Groups and Lie Algebras, Chapters 4–6*, Springer-Verlag, 2002.

[8] D. Bourguiba and S. Zarati, Depth and the Steenrod algebra, *Invent. Math.* **128** (1997), 589–602.

[9] M. P. Brodmann and R. Y. Sharp, *Local Cohomology: An Algebraic Introduction With Geometric Applications*, Cambridge University Press, 2007.

[10] A. Broer, The direct summand property in modular invariant theory, *Transform. Groups* **10** (2005), 5–27.

[11] W. Bruns and J. Herzog, *Cohen–Macaulay Rings*, Cambridge University Press, 1993.

[12] H. E. A. Campbell and I. P. Hughes, Vector invariants of $U_2(\mathbf{F}_p)$: A proof of a conjecture of Richman, *Adv. Math.* **126** (1997), 1–20.

[13] H. E. A. Campbell, R. J. Shank and D. L. Wehlau, Rings of invariants for modular representations of elementary abelian p-groups, *Transorm. Groups* **18** (2013), 1–22.

[14] H. E. A. Campbell and D. L. Wehlau, *Modular Invariant Theory*, Springer-Verlag, 2011.

[15] D. Derksen and G. Kemper, *Computational Invariant Theory*, Springer-Verlag, 2002.

[16] D. Eisenbud, *Commutative Algebra with a View Toward Algebraic Geometry*, Springer-Verlag, 1995.

[17] G. Ellingsrud and T. Skjelbred, Profondeur d'anneaux d'invariants en caractéristique p, *Compos. Math.* **41** (1980), 233–244.

[18] P. Fleischmann, Relative trace ideals and Cohen — Macaulay quotients of modular invariant rings. *Computational Methods for Representations of Groups and Algebras*, Birkhauser, 1999, 211–233.

[19] P. Fleischmann, The Noether bound in invariant theory of finite groups, *Adv. Math.* **156** (2000), 23–32.

[20] P. Fleischmann, G. Kemper and R. J. Shank, Depth and cohomological connectivity in modular invariant rings theory, *Tran. American Math. Soc.*, **357** (2005), 3605–3621.

[21] P. Fleischmann and R. J. Shank, The relative trace ideal and the depth of modular rings of invariants, *Archiv der Mathematik* **80** (2003), pp. 347–353.

[22] P. Fleischmann and C. F. Woodcock, Non-linear group actions with polynomial invariant rings and a structure theorem for modular Galois extensions, *Proc. London Math. Soc.* **103** (2011), 826–846.

[23] J. Fogarty, On Noether's bound for polynomial invariants of finite groups, *Electron. Res. Announc. Amer. Math. Soc.* **7** (2001), 5–7.

[24] H.-W. Henn, A variant of the proof of the Landweber Stong conjecture, *Proceedings of Symposia in Pure Mathematics* **63** (1998), *Amer. Math. Soc.* 271–275.

[25] M. Hochster and J. A. Eagon, Cohen Macaulay Rings, invariant theory and the generic perfection of determinantal loci, *Amer. J. Math.* **93** (1971), 1020–1058.

[26] D. B. Karagueuzian and P. Symonds, The module structure of a group action on a polynomial ring: a finiteness theorem. *J. Amer. Math. Soc.* **20** (2007), 931–967.

[27] G. Kemper, On the Cohen-Macaulay property of modular invariant rings, *J. Algebra* **215** (1999), 330–351.

[28] G. Kemper and G. Malle, The finite irreducible linear groups with polynomial ring of invariants, *Transformation Groups* **2** (1997), 57–89.

[29] H. Matsumura, *Commutative ring theory*, Cambridge University Press, 1986.

[30] H. Nakajima, Modular representations of abelian groups with regular rings of invariants, *Nagoya Math. J.* **86** (1982), 229–248.

[31] H. Nakajima, Regular rings of invariants for unipotent groups, *J. Algebra* **85** (1983), 253–286.

[32] M. D. Neusel and L. Smith, Invariant theory of finite groups, *Amer. Math. Soc.* 2002.

[33] D. R. Richman, On vector invariants over finite fields, *Adv. Math.* **81** (1990), 30–65.

[34] R. J. Shank and D. L. Wehlau, The transfer in modular invariant theory, *J. Pure Appl. Algebra* **142** (1999), 63–77.

[35] P. Symonds, On the Castelnuovo-Mumford regularity of rings of polynomial invariants, *Ann. Math.* **174** (2011), 499–517.

[36] P. Symonds, Group actions on rings and the Cech complex, *Advances in Math* **240** (2013), 291–301.

[37] C. A. Weibel, *An Introduction to Homological Algebra*, Cambridge University Press, 1994.

Chapter 5

Model Theory

Ivan Tomašić

School of Mathematical Sciences,
Queen Mary University of London, London E1 4NS, UK
i.tomasic@qmul.ac.uk

This chapter starts with a gentle introduction to model theory, aimed at applications in algebra. It then accelerates and turns into a survey of recent spectacular applications of model theory in diophantine geometry.

1. Introduction

Pure model theory is the study of mathematical structures via methods of mathematical logic. One studies models of (usually first-order) theories, and the interesting course is to start with assumptions of logical nature and investigate the consequences for the models, the ultimate goal being either a quantitative or qualitative classification of models.

From the point of view of applications, it is the study of *definability*. The goal is to describe the sets which can be defined using first-order formulae in a class of mathematical structures of interest, and pinpoint the properties that can be captured or approximated by such definable sets.

The aim of this chapter is to persuade the reader that model theory provides an important and powerful alternative point of view in many mainstream areas of pure mathematics. Model-theorists develop innovative "geometries", abandoning the strict confines of classical methods with ease (for example, in difference or differential fields), often with far-reaching outcomes.

We illustrate this amazing impact of model theory through a series of applications in algebra and number theory, from the initial uses of algebraically closed fields, through p-adic and motivic integration, all the way to model-theoretic proofs of major conjectures of diophantine geometry such as Manin–Mumford, Mordell–Lang and André–Oort conjectures.

2. First-Order Logic

2.1. *Syntax*

When we wish to study a certain class of mathematical structures using first-order logic, it is essential to choose a set of symbols suitable for expressing their properties. A *language* is a triple $\mathcal{L}$ consisting of:

- A set of function symbols $\{f_i : i \in I_0\}$ (where each f_i is of some arity n_i);
- A set of relation symbols $\{R_i : i \in I_1\}$ (where each R_i is of some arity m_i);
- A set of constant symbols $\{c_i : i \in I_2\}$.

The *syntax* determines which collections of symbols form admissible expressions in first-order logic. A precise inductive definition of such expressions is often a stumbling block in communication to non-logicians, so we choose to proceed informally, trusting the reader to have enough mathematical experience to recognise meaningful formulae.

A *first-order formula* in a language $\mathcal{L}$ (or an *$\mathcal{L}$-formula*) is any 'meaningful' *finite* string of symbols made out of:

- Symbols from $\mathcal{L}$;
- Equality symbol $=$;
- Variables $x_0, x_1, x_2, \ldots$;
- Logical connectives $\neg$, $\wedge$, $\vee$;
- Quantifiers $\exists$, $\forall$;
- Parentheses.

Example 1 (formulae in the language of groups). Let $\mathcal{L}$ be the language consisting of a single binary function (operation) "$\cdot$". Valid $\mathcal{L}$-formulae include:

- $\exists x_0 \forall x_1 \ \ x_0 \cdot x_1 = x_1 \wedge x_1 \cdot x_0 = x_1$.
- $\forall x_1 \ \ x_0 \cdot x_1 = x_1 \cdot x_0$.

Although $\mathcal{L}$ suffices for expressing all properties of groups, it is more economical to formulate group-theoretic discussions in the language $\mathcal{L}_g$

consisting of a binary operation $\cdot$, unary operation $^{-1}$ and a constant e. We may write:

- $\forall x_1 \quad x_1 \cdot e = x_1 \wedge e \cdot x_1 = x_1$;
- $\forall x_1 \quad x_1 \cdot x_1^{-1} = e \wedge x_1^{-1} \cdot x_1 = e$;
- $\exists x_1 \exists x_2 \quad x_0 = x_1 \cdot x_2 \cdot x_1^{-1} \cdot x_2^{-1}$.

Example 2 (formulae in the language of ordered rings). Let $\mathcal{L}_{or}$ be the language consisting of binary functions $+$, $\cdot$, a binary relation $<$ and constant symbols 0, 1. Valid $\mathcal{L}_{or}$-formulae include:

- $x_1 = x_2 \vee (\neg(x_1 < x_2))$;
- $\exists x_3 \quad x_1 + x_3 \cdot x_3 = x_2$;
- $\forall x_1 (x_1 = 0 \vee \exists x_2 \quad x_2 \cdot x_1 = 1)$;
- $x_1^2 + x_2^2 < 1 + 1 + 1 + 1$.

A variable is *bound* in a formula if it is in the scope of a quantifier, otherwise it is *free*. If variables $x_1, \ldots, x_n$ occur freely in a formula φ, we often express it by writing $\varphi(x_1, \ldots, x_n)$. A formula with no free variables is called a *sentence*.

2.2. *Semantics*

We now turn to the *semantics* of first-order logic, and discuss how to give meanings, or *interpretations,* of formulae in suitable mathematical structures.

Given a language $\mathcal{L} = (\{f_i\}, \{R_i\}, \{c_i\})$, an $\mathcal{L}$-*structure* is a set M in which each symbol from $\mathcal{L}$ is assigned an interpretation. In particular:

- For each n_i-ary function symbol f_i, we are given a function $f_i^M \colon M^{n_i} \to M$;
- For each m_i-ary relation symbol R_i, we are given a relation $R_i^M \subseteq M^{m_i}$;
- For each constant symbol c_i, we are given an element $c_i^M \in M$.

Let M be an $\mathcal{L}$-structure, $\varphi(x_1, \ldots, x_n)$ an $\mathcal{L}$-formula, and let $a_1, \ldots, a_n \in M$. We say that M *satisfies* $\varphi(a_1, \ldots, a_n)$ and write

$$M \models \varphi(a_1, \ldots, a_n),$$

if the property expressed by φ is true for $a_1, \ldots a_n$ within M, when all quantifiers are interpreted as ranging over elements of M. The *set of realisations* of $\varphi(x_1, \ldots, x_n)$ in M is the set

$$\varphi(M) = \{(a_1, \ldots, a_n) \in M^n \ : \ M \models \varphi(a_1, \ldots, a_n)\}.$$

Note, if φ is a sentence, it is either true or false in a given M. We say that
a set $X \subseteq M^n$ is *definable* (with parameters from a subset $B \subseteq M$) if there
is a formula $\varphi(x_1, \ldots, x_n, y_1, \ldots, y_m)$ and $b_1, \ldots, b_m \in B$ such that
$$X = \varphi(M, \bar{b}) = \{(a_1, \ldots, a_n) \ : \ M \models \varphi(a_1, \ldots, a_n, b_1, \ldots, b_m)\}.$$
Example 3 (Satisfaction). Consider $(\mathbb{N}, +, id, 0)$ and $(\mathbb{Z}, +, x \mapsto -x, 0)$
as $\mathcal{L}_g = (\cdot, ^{-1}, e)$-structures. Let φ be the sentence $\forall x_1 \exists x_2 \ \ x_1 \cdot x_2 = e$.
Then $\mathbb{N} \not\models \varphi$, while $\mathbb{Z} \models \varphi$.

Example 4 (Realisations). Consider the formulae:

- $\varphi(x_1) \equiv \exists x_2 \ \ x_1 = x_2 \cdot x_2$,
- $\varphi_4(x_5) \equiv \exists x_1 \exists x_2 \exists x_3 \exists x_4 \ \ x_5 = x_1 \cdot x_1 + x_2 \cdot x_2 + x_3 \cdot x_3 + x_4 \cdot x_4$,
- $\psi(x_1, x_2) \equiv x_1 \cdot x_1 + x_2 \cdot x_2 = 1$,

in the language $\mathcal{L}_r = (+, \cdot, 0, 1)$ of rings. Then:

- $\varphi(\mathbb{R}) = \mathbb{R}_0^+$;
- $\varphi(\mathbb{C}) = \mathbb{C}$;
- $\psi(\mathbb{R})$ is the unit circle in the 'plane' $\mathbb{R}^2$;
- $\varphi_4(\mathbb{Z}) = \mathbb{N}$ (by Lagrange's four square theorem).

Example 5 (Definability with and without parameters). Thinking
of $\mathbb{R}$ as an ordered ring, the set $\{x : x > \sqrt{2}\}$ is definable with no parameters
by the formula $\varphi(x) \equiv (1 + 1 < x \cdot x) \wedge (0 < x)$. On the other hand, in order
to define $\{x : x > \pi\}$, we need the parameter π.

Limitations of first-order logic stem from the stipulation that we can
only quantify over *elements* of a structure, and not over subsets, functions
or even natural numbers if they are not part of the structure. For example,
we *cannot*:

- Express that a group is torsion, see Exercise 1;
- Express that a graph is connected;
- Characterise $\mathbb{R}$ up to isomorphism (the Archimedean axiom is not first-order), see Exercise 4;
- State that a ring is a PID, etc.

2.3. *Morphisms of structures*

A map $f : M \to N$ between two $\mathcal{L}$-structures is a *homomorphism*, if:

- For each function symbol F of $\mathcal{L}$ of arity n, $f \circ F^M = F^N \circ f^n$;
- For each relation symbol R of $\mathcal{L}$ of arity m, $f^m(R^M) \subseteq R^N$;
- For each constant c of $\mathcal{L}$, $f(c^M) = c^N$.

An injective homomorphism is called an *embedding* if in addition $f^m(R^M) = R^N \cap f^m(M^m)$ for each relation symbol. An *isomorphism* is a surjective embedding $f : M \to N$. We write $M \cong N$ when M and N are isomorphic.

Given two $\mathcal{L}$-structures M and N with $N \subseteq M$, we say that N is a *substructure* of M if the inclusion $N \to M$ is an embedding. By abuse of notation, we write $N \subseteq M$ to express that N is a substructure of M.

Two $\mathcal{L}$-structures M and N are *elementarily equivalent*, written $M \equiv N$, if for every $\mathcal{L}$-sentence φ,

$$M \models \varphi \text{ if and only if } N \models \varphi.$$

An $\mathcal{L}$-embedding $f : M \to N$ is *elementary* if for all $a_1, \ldots, a_n \in M$ and for any formula $\varphi(x_1, \ldots, x_n)$,

$$M \models \varphi(a_1, \ldots, a_n) \text{ if and only if } N \models \varphi(f(a_1), \ldots, f(a_n)).$$

In other words, f preserves all formulae. If M is a substructure of N, we say that M is an *elementary substructure* if the inclusion map is elementary, and we write $M \preceq N$.

Example 6 (Non-elementary substructure). In the language of rings, $\mathbb{R}$ is a substructure of $\mathbb{C}$, since it is a subfield. On the other hand, in notation of Example 4, $\varphi(\mathbb{R}) = \mathbb{R}_0^+$ and $\varphi(\mathbb{C}) = \mathbb{C}$ so $\varphi(\mathbb{R}) \neq \varphi(\mathbb{C}) \cap \mathbb{R}$ and therefore $\mathbb{R} \not\preceq \mathbb{C}$.

Remark 1. A reader versed in category theory might prefer the following way of thinking about definable sets. Let $\mathcal{S}_\mathcal{L}$ be the category of $\mathcal{L}$-structures with elementary embeddings as morphisms. A formula $\varphi(\bar{x})$ can be identified with its realisation functor $F_\varphi : \mathcal{S}_\mathcal{L} \to \text{Set}$,

$$F_\varphi(M) = \varphi(M).$$

An implication $\varphi \to \psi$ induces a natural transformation $F_\varphi \to F_\psi$.

Proposition 1. *If M and N are isomorphic, then they are elementarily equivalent.*

In fact, by induction on the complexity of formulae, we can prove that an isomorphism $f : M \to N$ is an elementary embedding. Note that $N \preceq M$ implies $N \equiv M$, so any example of a proper elementary substructure will show that the converse fails, see Subsection 4.2.

2.4.   *Theories*

Definition 1.

(a) A *theory* in a language $\mathcal{L}$ is a set of $\mathcal{L}$-sentences (may be infinite).
(b) If T is a theory in a language $\mathcal{L}$, and M is an $\mathcal{L}$-structure, we say that M is a *model* of T, or that M models T, writing $M \models T$, if for every $\varphi \in T$, $M \models \varphi$.
(c) We say that φ is a *logical consequence* of T, $T \models \varphi$, if for every model M of T, $M \models \varphi$.
(d) We say that T *proves* φ, $T \vdash \varphi$, if there is a formal proof of the sentence φ starting from assumptions T.
(e) We say that T is *consistent* if it does not prove both φ and $\neg\varphi$ for some φ (a *contradiction*).

Example 7 (Theory of groups). The theory of groups consists of the usual group axioms, which happen to be first-order:

(a) $\forall x \forall y \forall z \ (x \cdot y) \cdot z = x \cdot (y \cdot z)$;
(b) $\forall x \ x \cdot e = x$;
(c) $\forall x \ x \cdot x^{-1} = e$.

Naturally, its models are groups.

Example 8 (Theory DLO of dense linear orders without endpoints). It consists of the following sentences in the language with a single binary relation $<$.

(a) $\forall x \ \neg(x < x)$;
(b) $\forall x \forall y (x = y \lor x < y \lor y < x)$;
(c) $\forall x \forall y \forall z (x < y \land y < z \to x < z)$;
(d) $\forall x \forall y (x < y \to \exists z (x < z \land z < y))$;
(e) $\forall x \exists z \ z < x$; $\forall x \exists z \ x < z$.

Example 9 (Theory ACF of algebraically closed fields). In first-order logic, we cannot quantify over integers, or polynomials of varying degrees. Thus, we require infinitely many axioms to state that a field is algebraically closed, as follows:

(1) The usual algebraic axioms for a field (which are first-order);

(2) Axiom schema: for every n, add the axiom
$$\forall y_0 \forall y_1 \ldots \forall y_{n-1} \exists x \quad y_0 + y_1 \cdot x + \cdots + y_{n-1} x^{n-1} + x^n = 0,$$
where we have used the notation x^k as a shorthand for $\underbrace{x \cdot x \cdot \ldots \cdot x}_{k \text{ times}}$.

Example 10 (Theory RCF of real closed ordered fields). It consists of these sentences:

(1) The usual axioms for ordered fields;
(2) $\forall x > 0 \ \exists y \ y^2 = x$;
(3) Axiom schema: for every odd n, take the axiom
$$\forall y_0 \cdots \forall y_{n-1} \exists x \quad x^n + y_{n-1} x^{n-1} + \cdots + y_0 = 0.$$

Example 11 (Peano arithmetic PA). The following axioms represent a first-order attempt to capture the essence of natural numbers. While the "standard model" of $\mathbb{N}$ is a model of this theory, it is too weak to exclude the "non-standard" models (Exercise 4), so mathematicians usually reach for a second-order characterisation of $\mathbb{N}$.

(1) $\forall x \ x + 1 \neq 0$;
(2) $\forall x \forall y (x + 1 = y + 1 \rightarrow x = y)$;
(3) $\forall x \ x + 0 = x$; $\forall x \forall y \ x + (y + 1) = (x + y) + 1$;
(4) $\forall x \ x \cdot 0 = 0$; $\forall x \forall y \ x \cdot (y + 1) = x \cdot y + x$;
(5) $\forall x \neg (x < 0)$; $\forall x \forall y (x < (y + 1) \leftrightarrow x < y \lor x = y)$;
(6) axiom schema: for each first-order formula $\varphi(x, \bar{z})$, include the axiom
$$\forall \bar{z} (\varphi(0, \bar{z}) \land \forall x (\varphi(x, \bar{z}) \rightarrow \varphi(x + 1, \bar{z})) \rightarrow \forall x \varphi(x, \bar{z})).$$

2.5. *Completeness and compactness*

The following result establishes a perfect correspondence between semantic truth and formal syntactic provability. It is a cornerstone of first-order logic and model theory.

Theorem 1 (Gödel's Completeness I). *For a theory T,*
$$T \models \varphi \text{ if and only if } T \vdash \varphi.$$

Theorem 2 (Completeness II). *For a theory T,*
$$T \text{ has a model if and only if } T \text{ is consistent.}$$

An easy consequence of the above is another fundamental tool of model theory.

Theorem 3 (Compactness Theorem). *If every finite subset of T has a model, then T has a model.*

Proof. If T has no model then, by Completeness II, T is inconsistent, i.e., there is a proof of a contradiction from T. Since proofs are finite sequences of statements, it can use only a finite number of assumptions from T, so there is a finite inconsistent subset of T, which in turn has no model by Completeness II.   □

An $\mathcal{L}$-theory T is *complete* if for every $\mathcal{L}$-sentence φ, either $T \models \varphi$ or $T \models \neg\varphi$. Equivalently, a theory is complete if any two of its models are elementarily equivalent.

The easiest way of producing a complete theory is to consider the *complete theory* of a given structure M,

$$\mathrm{Th}(M) = \{\varphi \ : \ M \models \varphi\}.$$

Example 12 ((In)complete theories).

(1) The theory of groups is not complete, since groups can be commutative and non-commutative.
(2) The theory of algebraically closed fields in not complete. It does not decide the characteristic.
(3) The theory of algebraically closed fields of fixed characteristic is complete, see Corollary 2.
(4) The theories DLO and RCF are complete, see Corollary 4 and Corollary 6.

The following celebrated result reveals inherent limitations of formal axiomatic approach to mathematics, defeating Hilbert's program to find a consistent and complete set of axioms for all of mathematics. In particular, it shows that Peano axioms are not complete and cannot be completed in a reasonable way.

Theorem 4 (Gödel's 1st Incompleteness Theorem). *For every consistent theory "containing" enough arithmetic, there is a statement which is true, but not provable in the theory.*

The idea of proof comes from the famous liar paradox, the self-referential sentence φ stating: "φ cannot be proved in T".

3. Basic Model Theory

3.1. *Size of models, categoricity, completeness, decidability*

Let us start with the following fundamental question. Given a consistent theory T, for which cardinals κ can we find models of size κ?

Theorem 5 (Down and up Löwenheim–Skolem). *Let $\mathcal{L}$ be a language.*

(1) *Let M be an $\mathcal{L}$-structure and $X \subseteq M$. Then there exists $M_0 \preceq M$ such that $X \subseteq M_0$ and $|M_0| \leq |X| + |\mathcal{L}|$.*
(2) *Let M be an infinite $\mathcal{L}$-structure. For any cardinal $\kappa > |M|$, M has an elementary extension of cardinality κ.*

Corollary 1. *If T is a countable theory with an infinite model, then T has models in all infinite cardinalities.*

We can now refine the initial question, and consider the number of models of a given size.

We say that a theory T is *categorical* in an infinite cardinality κ, or κ-*categorical*, if T has, up to isomorphism, a unique model of cardinality κ.

Theorem 6 (Vaught's test). *If all models of T are infinite and T is categorical in some infinite cardinality κ, then T is complete.*

Proof. Suppose T is not complete. Then there exists a sentence φ such that both $T \cup \{\varphi\}$ and $T \cup \{\neg\varphi\}$ are consistent. By Löwenheim–Skolem, we can find structures M and N of cardinality κ such that $M \models T \cup \{\varphi\}$ and $N \models T \cup \{\neg\varphi\}$. This is impossible, as M must be isomorphic to N. $\quad\square$

A theory T is *decidable* if there is an algorithm which determines for each sentence φ whether $T \models \varphi$. A theory T is *recursively enumerable* if there is an algorithm that lists the sentences of T.

Proposition 2. *A complete recursively enumerable theory is decidable.*

Proof. By completeness of T and the Completeness Theorem, there is either a proof of φ from T or a proof of $\neg\varphi$ from T. Thus, we can systematically search through all finite strings of symbols until we find a proof of either φ or $\neg\varphi$. $\quad\square$

3.2.   *The theory* ACF_p *and other examples.*

We shall now apply the model-theoretic tools of the previous subsection to study the **theory of algebraically closed fields of fixed characteristic.** For a prime number p, let ψ_p be the sentence

$$\forall x \quad \underbrace{x + x + \cdots + x}_{p \text{ times}} = 0.$$

- Let ACF_p be the theory of algebraically closed fields given in Example 9, together with ψ_p.
- Let ACF_0 be the theory of algebraically closed fields together with $\{\neg\psi_p : p \text{ prime}\}$.

Theorem 7. *The theory* ACF_p *is* κ-*categorical for every uncountable* κ.

Proof. Let p be either a prime or 0 and fix an uncountable cardinal κ. Take two algebraically closed fields K_1 and K_2 of characteristic p and cardinality κ with transcendence bases S_1 and S_2. Since K_i is the algebraic closure of $k(S_i)$ (where the prime subfield k is either $\mathbb{F}_p$ of $\mathbb{Q}$, depending on p), it follows that $|K_i| = |S_i| + \aleph_0$, so $|S_1| = |S_2|$. Pick any bijection $f : S_1 \to S_2$. It uniquely extends to an isomorphism $f : k(S_1) \to k(S_2)$, and, by the uniqueness of algebraic closure (up to isomorphism), we can find an isomorphism $K_1 \to K_2$.   □

By Vaught's test (Theorem 6), we get the following.

Corollary 2. *The theory* ACF_p *is complete for* $p \geq 0$.

Theorem 8 (The Lefschetz principle). *Let* φ *be a sentence in the language of rings. The following statements are equivalent:*

(1) φ *is true in* $\mathbb{C}$.
(2) φ *is true in every algebraically closed field of char 0.*
(3) φ *is true in some algebraically closed field of char 0.*
(4) *There are arbitrarily large primes* p *such that* φ *is true in some algebraically closed field of char* p.
(5) *There is an* m *such that for all* $p > m$, φ *is true in all algebraically closed fields of characteristic* p.

Proof. The equivalence of 1–3 is just the completeness of ACF_0 and $5 \Rightarrow 4$ is obvious.

For $2 \Rightarrow 5$, suppose $\mathrm{ACF}_0 \models \varphi$. By the completeness theorem, $\mathrm{ACF}_0 \vdash \varphi$ and the proof uses only finitely many $\neg\psi_p$. Thus, for large enough p, $\mathrm{ACF}_p \models \varphi$.

For $4 \Rightarrow 2$, suppose $\mathrm{ACF}_0 \not\models \varphi$. By completeness $\mathrm{ACF}_0 \models \neg\varphi$. By the above argument, $\mathrm{ACF}_p \models \neg\varphi$ so 4 fails. $\qquad\square$

The following is a curious application of the Lefschetz principle.

Theorem 9 (Ax/Grothendieck). *Let $f : \mathbb{C}^n \to \mathbb{C}^n$ be an injective polynomial map. Then f is surjective.*

Proof. Let $f(\bar{x}) = (f_1(\bar{x}), \ldots, f_n(\bar{x}))$ and suppose $f_i \in \mathbb{C}[\bar{x}]$ are of total degrees less than some d. Let $\phi_{n,d}$ be the first-order sentence stating that every injective polynomial map in n coordinates whose coordinate functions are of degree at most d is surjective.

Clearly, if k is any finite field, $k \models \phi_{n,d}$, and, since $\phi_{n,d}$ ultimately has the form of an $\forall\exists$-statement, the same is true for any increasing union of finite fields (see Exercise 5). In particular, the algebraic closure of any finite field satisfies $\phi_{n,d}$ and the Lefschetz principle implies that $\phi_{n,d}$ also holds for $\mathbb{C}$. $\qquad\square$

Corollary 3. *The theory ACF_p is decidable for each $p \geq 0$.*

We can adopt a similar approach to study dense linear orders without endpoints (Example 8), by virtue of the following classical result.

Theorem 10 (Cantor). *If A and B are countable dense linear orders without endpoints, then $A \cong B$.*

Proof. We use a *back and forth* argument. Suppose we have a partial order-preserving bijection $f : A_0 \to B_0$, where A_0 is a finite subset of A, and B_0 is a finite subset of B. In the "forth" direction, it is clearly possible, given $a \in A \setminus A_0$ to find some $b \in B \setminus B_0$ such that $f \cup \{(a,b)\}$ is again a partial order-preserving bijection. In the "back" direction, given $b \in B \setminus B_0$, we can find $a \in A \setminus A_0$ such that $f \cup \{(a,b)\}$ is again a partial order-preserving bijection.

Enumerate A as a_i, $i \in \omega$, and B as b_i. We inductively form a sequence $f_j : A_j \to B_j$ of order-preserving partial bijections such that for every i, $a_i \in A_{2i} = \mathrm{Dom}\, f_{2i}$ and $b_i \in B_{2i+1} = \mathrm{Im}\, f_{2i+1}$. The map $\cup_j f_j$ is an isomorphism. $\qquad\square$

Cantor's theorem in model-theoretic terms states that the theory DLO is $\aleph_0$-categorical, and we can invoke Vaught's test.

Corollary 4. *The theory* DLO *is complete.*

Note that, even in the case of algebraically closed fields, some of the back-and-forth machinery appears disguised in the proof of uniqueness of the algebraic closure (up to isomorphism).

3.3. *Types*

Let M be an $\mathcal{L}$-structure and let $A \subseteq M$. Let $\mathcal{L}_A$ be the language obtained by adding to $\mathcal{L}$ the constant symbols for all elements of A. Let $\mathrm{Th}_A(M)$ be the set of all $\mathcal{L}_A$-sentences true in M.

Definition 2.

(1) An *n-type over A* is a set of $\mathcal{L}_A$-formulae in free variables $x_1, \ldots, x_n$ that is consistent with $\mathrm{Th}_A(M)$.
(2) An n-type p over A is *complete* if for every $\mathcal{L}_A$-formula $\varphi(\bar{x})$, either $\varphi(\bar{x}) \in p$ or $\neg\varphi(\bar{x}) \in p$.
(3) Incomplete types are sometimes called *partial*.
(4) Let $S_n(A)$ denote the set of all complete n-types over A.

An easy way of producing a complete type is the following. Let N be an elementary extension of M and let $\bar{b} \in N^n$. Then

$$tp(\bar{b}/A) = \{\varphi(\bar{x}) \in \mathcal{L}_A \ : \ N \models \varphi(\bar{b})\}$$

is a complete type.

On the other hand, if we have a partial n-type over A, it will be realised in some elementary extension N of M.

Proposition 3. *Let $\bar{a}, \bar{b} \in M^n$ and $tp(\bar{a}/A) = tp(\bar{b}/A)$. Then there is an elementary extension N of M and an $\mathcal{L}$-automorphism of N fixing A and mapping $\bar{a}$ to $\bar{b}$.*

Proof. We iterate the following lemma. Suppose M is an $\mathcal{L}$-structure, $A \subseteq M$ and $f : A \to M$ is a partial elementary map, i.e., $M \models \varphi(a_1, \ldots, a_n)$ if and only if $M \models \varphi(f(a_1), \ldots, f(a_n))$ for all $a_i \in A$. If $b \in M$, we can find an elementary extension N of M and extend f to a partial elementary map from $A \cup \{b\}$ into N. $\square$

For each $\mathcal{L}_A$-formula $\varphi(x_1, \ldots, x_n)$, consider the set

$$B_\varphi = \{p \in S_n(A) \ : \ \varphi \in p\}.$$

The *Stone topology* on $S_n(A)$ is generated by the basic open sets B_φ.

Theorem 11. *The* Stone space $S_n(A)$ *is compact and totally disconnected.*

Proof. Note that $S_n(A) \setminus B_\varphi = B_{\neg\varphi}$ so each B_φ is open and closed. Thus $S_n(A)$ is totally disconnected. Suppose $\{B_{\varphi_i} : i \in I\}$ is a cover of $S_n(A)$ by basic open sets. Then

$$\bigcap_i B_{\neg\varphi_i} = \emptyset$$

and thus the set $\{\neg\varphi_i : i \in I\}$ is inconsistent. By Compactness, it has a finite inconsistent subset $\{\neg\varphi_i : i \in I_0\}$ and $\{B_{\varphi_i} : i \in I_0\}$ is a finite subcover. $\square$

In the above discussion of types, in order to find realisations, or to relate types to Galois considerations, we had to go to elementary extensions. A desire to do this without changing the model leads to the notion of saturation.

Let κ be an infinite cardinal. We say that a structure M is κ-*saturated* if for every $A \subseteq M$ with $|A| < \kappa$, all the types in $S_1(A)$ are realised in M. We say that M is *saturated* if it is $|M|$-saturated.

An easy inductive argument shows that if M is κ-saturated and $|A| < \kappa$, then every type in $S_n(A)$ is realised in M^n. We present some advantages of working with saturated models.

Proposition 4. *Suppose M is saturated. Then:*

(1) *M is* strongly homogeneous, *i.e., if $A \subseteq M$ and $|A| < |M|$, then $tp(\bar{a}/A) = tp(\bar{b}/A)$ if and only if there is an automorphism of M fixing A and mapping $\bar{a}$ to $\bar{b}$ (complete types over A are $\mathrm{Aut}(M/A)$-orbits).*
(2) *M is* universal, *i.e., every small model of $\mathrm{Th}(M)$ embeds into M.*

Proposition 5 (ACF and saturation). *An algebraically closed field K is saturated if and only if it is of infinite transcendence degree.*

Proof. Suppose $A \subset K$ is finite and F is the field generated by A. Let p be the 1-type over A which states that x is transcendental over F. If K is $\aleph_0$-saturated, then p must be realised in K. Thus, we conclude by induction that an $\aleph_0$-saturated algebraically closed field must be of infinite transcendence degree.

Conversely, suppose K has infinite transcendence degree and $F \subseteq K$ is a field generated by fewer than $|K|$ elements. Consider the ideal

$$I_p := \{f(x) \in F[x] : \text{`} f(x) = 0\text{'} \in p\}.$$

If $I_p = 0$, then p says that x is transcendental over F, and we can find a realisation in K. If $I_p \neq 0$, since $F[x]$ is a PID, I_p is generated by some $f(x)$ and any zero of f in K realises p. □

A subset of a saturated model is *type-definable*, if it is the set of realisations of a partial type. In this way, we can deal with infinite conjunctions of formulae without the need for higher-order logics.

In general, the upper bound for the number of types over a set A is

$$|S_n(A)| \leq 2^{|A|+|\mathcal{L}|+\aleph_0},$$

which is sometimes attained, so there are set-theoretic problems associated with finding saturated models. Typically we need to assume the generalised continuum hypothesis or the existence of inaccessible cardinals to be able to find them. Saturated models exist unconditionally if the theory is *stable*, as discussed in Subsection 5.3.

4. Applications in Algebra

4.1. *Quantifier elimination*

We say that a theory T admits *quantifier elimination (QE)* if for every formula $\psi(\bar{x})$ there is a quantifier-free formula $\varphi(\bar{x})$ such that

$$T \models \forall \bar{x} \ (\psi(\bar{x}) \leftrightarrow \varphi(\bar{x})).$$

Theorem 12. *The theory* DLO *has QE.*

Proof. Let us write $x \leq y$ as a shorthand for $(x < y) \vee (x = y)$. After some thinking, one sees that every quantifier-free formula is a $\wedge, \vee$-combination of formulae of form $x \leq y$ and $x < y$. Since $\exists x(\varphi \vee \psi)$ is equivalent to $(\exists x \varphi) \vee (\exists x \psi)$, it is enough to show how to eliminate the quantifier from

$$\exists x \bigwedge_i (y_i < x) \wedge \bigwedge_j (y_j \leq x) \wedge \bigwedge_k (x \leq z_k) \wedge \bigwedge_l (x < z_l),$$

but this is clearly equivalent to

$$\bigwedge_{i,k}(y_i < z_k) \wedge \bigwedge_{i,l}(y_i < z_l) \wedge \bigwedge_{j,l}(y_j < z_l) \wedge \bigwedge_{j,k}(y_j \leq z_k).$$

□

Proposition 6 (QE test). *Assume the language $\mathcal{L}$ contains at least one constant symbol. Let T be an $\mathcal{L}$-theory and let $\varphi(\bar{x})$ be an $\mathcal{L}$-formula (we allow φ to be a sentence). The following statements are equivalent.*

(1) *There is a quantifier-free formula $\psi(\bar{x})$ such that*

$$T \models \forall \bar{x} \ (\varphi(\bar{x}) \leftrightarrow \psi(\bar{x})).$$

(2) *If A and B are models of T and C a common substructure of A and B, then $A \models \varphi(\bar{a})$ if and only if $B \models \varphi(\bar{a})$ for all $\bar{a} \in C$.*

Proof. The direction $1 \Rightarrow 2$ is trivial, since quantifier-free formulae are preserved under (substructure) embeddings.

For $2 \Rightarrow 1$, if $T \models \forall \bar{x} \ \varphi(\bar{x})$, then $T \models \forall \bar{x} \ (\varphi(\bar{x}) \leftrightarrow c = c)$. If $T \models \forall \bar{x} \ \neg\varphi(\bar{x})$, then $T \models \forall \bar{x} \ (\varphi(\bar{x}) \leftrightarrow c \neq c)$. Thus we may assume that both $\varphi(\bar{x})$ and $\neg\varphi(\bar{x})$ are consistent with T. Define

$$\Gamma(\bar{x}) = \{\psi(\bar{x}) : \psi \text{ is quantifier-free and } T \models \forall \bar{x}(\varphi(\bar{x}) \rightarrow \psi(\bar{x}))\}.$$

Let $\bar{d}$ be new constant symbols. We will show below that $T \cup \Gamma(\bar{d}) \models \varphi(\bar{d})$. Assuming this, by compactness there are $\psi_1, \ldots, \psi_n \in \Gamma$ such that $T \models \forall \bar{x}(\bigwedge_i \psi_i(\bar{x}) \rightarrow \varphi(\bar{x}))$. By definition of Γ, we get $T \models \forall \bar{x}(\bigwedge_i \psi_i(\bar{x}) \leftrightarrow \varphi(\bar{x}))$ and $\bigwedge_i \psi_i(\bar{x})$ is quantifier-free.

It remains to prove

$$T \cup \Gamma(\bar{d}) \models \varphi(\bar{d}).$$

If not, let $A \models T \cup \Gamma(\bar{d}) \cup \{\neg\varphi(\bar{d})\}$. Let C be the substructure of A generated by $\bar{d}$ (if φ is a sentence the constant symbol ensures C non-empty). Let $\text{Diag}(C)$ be the set of all atomic and negated atomic formulas with parameters from C that are true in C. Let $\Sigma = T \cup \text{Diag}(C) \cup \varphi(\bar{d})$. If Σ is inconsistent, since C is generated by $\bar{d}$, there are quantifier-free formulae $\psi_1(\bar{d}), \ldots, \psi_n(\bar{d}) \in \text{Diag}(C)$ such that $T \models \forall \bar{x}(\bigwedge_i \psi_i(\bar{x}) \rightarrow \neg\varphi(\bar{x}))$. But then $T \models \forall \bar{x}(\varphi(\bar{x}) \rightarrow \bigvee_i \neg\psi_i(\bar{x}))$. So $\bigvee_i \neg\psi_i(\bar{x}) \in \Gamma$ and $C \models \bigvee_i \neg\psi_i(\bar{x})$, a contradiction. Thus Σ is consistent.

Let $B \models \Sigma$. Since $\text{Diag}(C) \subseteq \Sigma$, C embeds in B. But, since $A \models \neg\varphi(\bar{d})$, $B \models \neg\varphi(\bar{d})$, a contradiction. □

Lemma 1. *Suppose that for every quantifier-free formula $\theta(x,\bar{y})$ there is a quantifier-free formula $\psi(\bar{y})$ such that*

$$T \models \forall \bar{y}\, (\exists x \theta(x,\bar{y}) \leftrightarrow \psi(\bar{y})).$$

Then T has QE.

Proof. By induction on complexity of φ. Everything is trivial if φ is quantifier-free. The induction step is straightforward when φ is a Boolean combination of formulae for which QE works. If $\varphi(\bar{y}) \equiv \exists x \theta(x,\bar{y})$, by inductive hypothesis, we first find a quantifier-free formula $\psi_0(x,\bar{y})$ equivalent to $\theta(x,\bar{y})$, and then by assumption a quantifier-free formula $\psi(\bar{y})$ equivalent to $\exists x \psi_0(x,\bar{y})$. Then clearly $\psi(\bar{y})$ is a quantifier-free formula equivalent to $\varphi(\bar{y})$. $\qquad\square$

This means that it suffices to check the condition in the QE test for $\exists_1$-formulae.

Theorem 13 (Tarski). *The theory ACF has QE.*

Essentially, the same result was proved independently in algebraic geometry as the following restatement. Recall that a *constructible* subset of an algebraic variety X is a Boolean combination of Zariski closed subsets (a finite union of locally closed sets), and it can therefore be identified with a quantifier-free formula.

Theorem 14 (Chevalley). *Let $f : X \to Y$ be a morphism of finite type. The image of a constructible subset of X under f is constructible in Y.*

Proof of Theorem 13. Let F be a field and let K, L be algebraically closed extensions of F, and let $\bar{F}$ be the algebraic closure of F, viewed as a subfield of both K and L. Let $\varphi(x,\bar{y})$ be a quantifier-free formula, $a \in K$, $\bar{b} \in F$ such that $K \models \varphi(a,\bar{b})$. We need to show $L \models \exists x \varphi(x,\bar{b})$.

There are polynomials $f_{ij}, g_{ij} \in F[x]$ such that $\varphi(x,\bar{b})$ is equivalent to

$$\bigvee_i \left(\bigwedge_j f_{ij}(x) = 0 \ \wedge\ \bigwedge_j g_{ij}(x) \neq 0 \right).$$

Then $K \models \bigwedge_j f_{ij}(a) = 0 \ \wedge\ \bigwedge_j g_{ij}(a) \neq 0$ for some i. If not all f_{ij} are identically zero for that i, then $a \in \bar{F} \subseteq L$ and we are done. Otherwise, since $\bigwedge_j g_{ij}(a) \neq 0$, all g_{ij} are non-zero polynomials and have finitely many roots in L, so we can easily find an element $d \in L$ which satisfies all inequations, and $L \models \varphi(d,\bar{b})$. $\qquad\square$

Let us discuss some applications of QE for ACF.

Corollary 5. *Let K be an algebraically closed field.*

(1) *Any definable subset of K in one variable is either finite or cofinite (this is the property we call* strong minimality*).*
(2) *Let $f : K \to K$ be a definable function. If K is of characteristic 0, there is a rational function $g \in K(x)$ such that $f(a) = g(a)$ for all but finitely many $a \in K$. If K is of characteristic p, there is a rational function g and $n \geq 0$ such that $f(a) = g(a)^{1/p^n}$ for almost all $a \in K$.*

Proof. (1) Every definable set in one variable is a Boolean combination of sets of the form $\{x : f(x) = 0\}$ for a polynomial f and such sets are finite.

Statement (2) is proved by a "generic point" argument. Assume $\mathrm{char}(K) = 0$. Let L be a proper elementary extension of K and let $a \in L \backslash K$. If σ is any automorphism of L fixing $K(a)$, then $\sigma(f(a)) = f^\sigma(\sigma(a)) = f(a)$. We conclude $f(a) \in K(a)$ and thus there is a rational function $g \in K(x)$ such that $f(a) = g(a)$. Consider $\varphi(x) \equiv f(x) = g(x)$. By above, $\varphi(K)$ is either finite or cofinite. If it were of size N, since $K \preceq L$, $\varphi(L)$ would also be of size N, contradicting the fact that $a \in \varphi(L) \setminus \varphi(K)$. Thus $\varphi(K)$ must be cofinite. In characteristic p, we have to be mindful of the Frobenius morphism $x \mapsto x^p$, whose inverse is definable but not algebraic. $\square$

For readers with algebraic geometry background, we give a comparison of the Stone space and the prime spectrum over algebraically closed fields. Extending the notation from the proof of Proposition 5, we write $I_p := \{f \in F[x_1, \ldots, x_n] : `f(x_1, \ldots, x_n) = 0' \in p\}$, for $p \in S_n(F)$.

Proposition 7. *If K is algebraically closed and F a subfield of K, the map $p \mapsto I_p$ is a continuous bijection between $S_n(F)$ and $\mathrm{Spec}(F[x_1, \ldots, x_n])$.*

Proof. If $fg \in I_p$, then $f(\bar{x})g(\bar{x}) = 0 \in p$. Since p is complete, either $f(x) = 0 \in p$ or $g(x) = 0 \in p$, so I_p is prime. If $\mathfrak{p}$ is a prime ideal, we can find a prime ideal $\mathfrak{p}_1$ in $K[\bar{x}]$ such that $\mathfrak{p}_1 \cap F[\bar{x}] = \mathfrak{p}$. Let K_1 be the algebraic closure of $K[\bar{x}]/\mathfrak{p}_1$ and let $a_i = x_i + \mathfrak{p}_1$. For $f \in K[\bar{x}]$, $f(\bar{a}) = 0$ if and only if $f \in \mathfrak{p}_1$ and thus $I_{\mathrm{tp}(\bar{a}/F)} = \mathfrak{p}$ so the map is surjective. By QE, if $p \neq q$, then $I_p \neq I_q$, so the map is injective. Continuity is obvious. $\square$

We return to the theory of real closed fields from Example 10.

Theorem 15 (Tarski). *The theory* RCF *has QE in the language of ordered rings.*

Proof. We use the QE test. Let F_0, F_1 be models of RCF and let R be a common substructure (an ordered domain). Let L be the real closure of the fraction field of R. We may assume L is a substructure of F_0 and F_1. Suppose $\varphi(x, \bar{y})$ is quantifier-free, $\bar{b} \in R$, $a \in F_0$ and $F_0 \models \varphi(a, \bar{b})$. We need to show $F_1 \models \exists x\, \varphi(x, \bar{y})$, but it is enough to show $L \models \exists x\, \varphi(x, \bar{y})$. We can find polynomials $f_i, g_j \in R[x]$ so that $\varphi(x, \bar{b})$ is equivalent to

$$\bigwedge_i f_i(x) = 0 \ \wedge \ \bigwedge_j g_j(x) > 0.$$

If not all of f_i are zero, it follows that a is algebraic over R and thus in L and we are done. Thus, we reduce to the case $\varphi(x, \bar{b}) \equiv \bigwedge_j g_j(x) > 0$. Since L is real closed, we can factor each g_j as a product of factors of form $(x - c)$ or $x^2 + bx + c$ with $b^2 - 4c < 0$. Linear factors change sign at c, quadratic factors do not change sign, so we are left with a (consistent) linear system of inequalities which can definitely be solved in L. □

Corollary 6. *The theory* RCF *is complete and decidable.*

Proof. We can embed $\mathbb{Q}$ inside any real closed field F. Given a sentence φ, find the equivalent quantifier-free formula ψ, and we get $F \models \varphi$ if and only if $\mathbb{Q} \models \psi$. Thus, for any two real closed fields F_1 and F_2, $F_1 \models \varphi$ if and only if $\mathbb{Q} \models \psi$ if and only if $F_2 \models \varphi$. Decidability is a direct consequence of completeness and Proposition 2. □

Corollary 7. *A definable subset (in one variable) of a real closed field is a finite union of points and intervals.*

An abstract structure with the above property is called *o-minimal*. Recently, o-minimality has found impressive applications in the work surrounding the André–Oort conjecture, see Section 8.

Proof. Definable sets in one variable are Boolean combinations of $\{x : f(x) > 0\}$ which are finite unions of intervals. □

4.2. Model completeness

A theory T is *model complete* if $M, N \models T$ and $M \subseteq N$ implies $M \preceq N$.

Proposition 8. *If T has QE, then T is model complete.*

Proof. Let $M \subseteq N$ be models of T. Suppose $\varphi(\bar{x})$ is a formula and $\bar{a} \in M$. There is a quantifier-free formula $\psi(\bar{x})$ such that $T \models \forall \bar{x}\ (\varphi(\bar{x}) \leftrightarrow \psi(\bar{x}))$.

Since $\psi(\bar{x})$ is quantifier-free, $M \models \psi(\bar{a})$ if and only if $N \models \psi(\bar{a})$. Thus, $M \models \varphi(\bar{a})$ if and only if $N \models \varphi(\bar{a})$. $\qquad\square$

Earlier QE results immediately yield that the theories DLO, ACF, RCF are model-complete. On the other hand, a theory can be model complete without admitting QE. For example:

- Th($\mathbb{R}$) in the language of rings.
- Wilkie's example: Th($\mathbb{R}, +, -, \cdot, <, \exp, 0, 1$) is model-complete, yet one cannot eliminate the quantifier from

$$y > 0 \wedge \exists w \ (wy = x \ \wedge \ z = y \exp(w)).$$

We use model-completeness of ACF to give a rather short proof of Hilbert's Nullstellensatz.

Theorem 16 (Weak Nullstellensatz). *Let F be an algebraically closed field and let $I \subseteq F[x_1, \ldots, x_n]$ be a prime ideal. Then there is $\bar{a} \in F^n$ such that $f(\bar{a}) = 0$ for all $f \in I$.*

Proof. Let K be the algebraic closure of the fraction field of $F[x_1, \ldots, x_n]/I$. If we denote $b_i = x_i + I$, then $f(b_1, \ldots, b_n) = 0$ for all $f \in I$. If $f_1, \ldots, f_m$ generate I, then

$$K \models \exists \bar{y} \ \bigwedge_i f_i(\bar{y}) = 0.$$

By model completeness, this sentence is already true in F. $\qquad\square$

The full version of Nullstellensatz, $I(V(J)) = \sqrt{J}$, is easily obtained from the weak one using the "Rabinowitsch trick", see [8, IX.1.5].

There is also a neat solution to Hilbert's 17th problem using model completeness of RCF.

Theorem 17 (Artin). *Let F be a real closed field. Suppose that $f \in F(x_1, \ldots, x_n)$ is such that $f(\bar{a}) \geq 0$ for all $\bar{a} \in F^n$. Then f is a sum of squares of rational functions.*

Proof. If not, we can extend the order of F to $F(x_1, \ldots, x_n)$ so that $f < 0$. Let K be the real closure of $F(x_1, \ldots, x_n)$ with this ordering. Then

$$K \models \exists \bar{y} \ f(\bar{y}) < 0,$$

since $K \models f(\bar{x}) < 0$. By model completeness, F satisfies the same sentence, which is a contradiction. $\qquad\square$

4.3.   *Valued fields*

A *valuation* of a field K with values in a totally ordered Abelian group Γ is a mapping $v : K^\times \to \Gamma$, such that:

(1) $v(xy) = v(x) + v(y)$,
(2) $v(x + y) \geq \min(v(x), v(y))$

for all $x, y \in K^\times$. The set $R = \{0\} \cup \{x \in K^\times : v(x) \geq 0\}$ is a local ring called the *valuation ring* whose unique maximal ideal is $\mathfrak{m} = \{0\} \cup \{x \in K^\times : v(x) > 0\}$, and the *residue field* is $k = R/\mathfrak{m}$.

A valued field K is *henselian* if, for every polynomial $f \in R[t]$, writing $\bar{f}$ for the reduction of f in $k[t]$, every simple root $\bar{x} \in k$ of $\bar{f}$ has a unique lifting $x \in R$ to a root of f.

A *p-adically closed field* is a henselian valued field K of characteristic 0, such that the unique maximal ideal in the valuation ring R is pR, the residue field is $k = \mathbb{F}_p$ and the valuation group Γ satisfies $[\Gamma : n\Gamma] = n$. In a suitable language (e.g., with *sorts* for the field K, ring R, ordered group Γ, and function symbols for the valuation v and the residue map $R \to k$), all these requirements are first–order, and we can study the *theory of p-adically closed fields*.

Example 13 (*p*-adic numbers). Let p be a prime.

(1) The ring of *p-adic integers* is $\mathbb{Z}_p = \varprojlim_m \mathbb{Z}/p^m\mathbb{Z}$.
(2) The field of *p-adic numbers* $\mathbb{Q}_p$ is the fraction field of $\mathbb{Z}_p$.

An equivalent description can be obtained through norm and completion. For non-zero $a, b \in \mathbb{Z}$, the rules $v_p(a) = \max\{n : p^n \text{ divides } a\}$ and $v_p(\frac{a}{b}) = v_p(a) - v_p(b)$ define the *p-adic valuation* on $\mathbb{Q}$. Then $\mathbb{Q}_p$ is the completion of $\mathbb{Q}$ with respect to the *p*-adic norm $|x|_p = p^{-v_p(x)}$. Its value ring is $\mathbb{Z}_p$, whose unique maximal ideal is $\mathfrak{p} = p\mathbb{Z}_p$, and the residue field is $\mathbb{Z}_p/\mathfrak{p} = \mathbb{F}_p$.

Theorem 18 (Hensel's Lemma). *The valued field $\mathbb{Q}_p$ is henselian.*

The above considerations show that $\mathbb{Q}_p$ is a *p*-adically closed field.

Theorem 19 (Ax–Kochen). *The theory of p-adically closed fields is complete and model complete.*

Theorem 20 (Ax–Kochen/Ershov). *Let K be a henselian valued field of equicharacteristic 0 (i.e., $\mathrm{char}(k) = \mathrm{char}(K) = 0$). Then $\mathrm{Th}(K)$ is determined by $\mathrm{Th}(k)$ and $\mathrm{Th}(\Gamma)$.*

Corollary 8. *Let φ be a first-order sentence about valued fields. Then*

$$\mathbb{Q}_p \models \varphi \text{ if and only if } \mathbb{F}_p((t)) \models \varphi$$

for all but finitely many primes p.

Theorem 21 (Macintyre). *The theory of p-adically closed fields admits QE in the language of valued fields extended with predicates P_n for n-th powers.*

4.4. Further topics: Poincaré series, motivic integration

Let $f(x) \in \mathbb{Z}[x]$ be a polynomial in variables $x = (x_1, \dots, x_n)$ and fix a prime p. Let:

- $N_m = |\{x \in (\mathbb{Z}/p^m\mathbb{Z})^n : f(x) \equiv 0 \mod p^m\}|$,
- $\tilde{N}_m = |\{x \mod p^m : x \in \mathbb{Z}_p^n, \ f(x) = 0\}|$.

The associated *Poincaré series* are

$$P_p(T) = \sum_m N_m T^m, \quad \tilde{P}_p(T) = \sum_m \tilde{N}_m T^m,$$

Igusa proved that $P(T)$ is a rational function of T using Hironaka's resolution of singularities, and Denef proved $\tilde{P}(T)$ is rational using resolution and Macintyre's QE. He also gave a proof of rationality of more general series using *cell decomposition*.

The main idea of Igusa/Denef proofs is to replace the counting of solutions of congruences by the calculation of certain p-adic integrals, where one can use familiar tools for integration such as the substitution of variables and Fubini's Theorem.

Recall, $\mathbb{Q}_p$ with addition is a locally compact group (and $\mathbb{Z}_p$ is compact), so there exists a unique Haar measure $\mu = |dx|$ on $\mathbb{Q}_p^n$ normalised so that $\mathbb{Z}_p^n$ has measure 1. Since the measure is translation-invariant, the measure of any coset $a + p^m\mathbb{Z}_p^n$, $a \in \mathbb{Q}_p^n$, is p^{-mn} and thus

$$N_m = p^{mn} \cdot \mu\{x \in \mathbb{Z}_p^n : v_p(f(x)) \geq m\}.$$

Igusa's local zeta function is defined as the p-adic integral

$$Z_{f,p}(s) = Z(s) = \int_{\mathbb{Z}_p^n} |f(x)|_p^s |dx|$$

for $s \in \mathbb{C}$, $\mathrm{Re}(s) \geq 0$.

It is easy to check that

$$P(p^{-n-s}) = \frac{1 - p^{-s} Z(s)}{1 - p^{-s}}$$

and thus the rationality of $P(T)$ is equivalent to $Z(s)$ being a rational function of p^{-s}.

On the other hand,

$$\tilde{N}_m = p^{mn} \cdot \mu\{x \in \mathbb{Z}_p^n : \exists y \in \mathbb{Z}_p^n, \, f(y) = 0, \, y \equiv x \mod p^m\},$$

the set on the right being much more complicated than the one appearing in the expression for N_m. Thus, Denef must expand the class of functions he is integrating and realises that the smallest class of functions closed under all necessary operations is that of definable functions in a suitable language. He now proves that, for a definable set S contained in a compact subset of $\mathbb{Q}_p^n$ and for a definable function $h : \mathbb{Q}_p^n \to \mathbb{Q}_p$ whose norm is bounded on S, the integral

$$Z(s) = \int_S |h(x)|_p^s |dx|$$

is a rational function of p^{-s}.

Towards an even deeper understanding of the subject, Macintyre noticed that the degrees of the numerator and the denominator in the expression of $P_p(T)$ as a rational function are bounded independently of p, which led to a speculation that there may be a "universal" Poincaré series which specialises to P_p for almost all primes p.

Denef and Loeser developed the theory of arithmetic motivic integration to explain (given $f \in \mathbb{Z}[x_1, \ldots, x_n]$) the uniformity in the expression of p-adic integrals $Z_{f,p}(s)$ as rational functions of p^{-s} when p varies. By virtue of QE results for valued fields which essentially push all the difficulty to the residue field, it turns out that the sought-after universal rational function must be expressed using coefficients in the Grothendieck ring of definable sets over finite fields (or the category of Chow motives).

5.   Dimension, Rank, Stability

5.1.   *Standard assumptions*

We consider a *complete* theory T in a *countable* language $\mathcal{L}$ with infinite models, and henceforth we fix a large saturated model $\mathfrak{C}$, sometimes called the *monster model*. Thus, by Proposition 4, it is homogeneous and any (small) model M of T is an elementary substructure of $\mathfrak{C}$.

For $A \subseteq \mathfrak{C}$ (small), the *definable closure* of A is

$$\mathrm{dcl}(A) = \{c \in \mathfrak{C} : c \text{ is fixed by any automorphism of } \mathfrak{C} \text{ fixing } A\},$$

whereas the *algebraic closure* of A is

$$\mathrm{acl}(A) = \{c \in \mathfrak{C} : c \text{ has finitely many conjugates over } A\}.$$

Clearly, $c \in \mathrm{dcl}(A)$ if and only if $\{c\}$ is an A-definable set, and $c \in \mathrm{acl}(A)$ if and only if c is contained in a finite A-definable set.

Example 14. Let $A \subseteq \mathfrak{C} \models \mathrm{ACF}$. Then $\mathrm{dcl}(A)$ is the perfect closure of the field generated by A, and $\mathrm{acl}(A)$ is the (usual) algebraic closure of the field generated by A. In models of DLO, acl is trivial.

5.2. *Morley rank*

Model theory considers a variety of rank functions, which, intuitively speaking, generalise the classical dimension notions, but sometimes take values in the class On of ordinals.

Definition 3. To a definable set X we associate its $\mathrm{On} \cup \{-1, \infty\}$-valued *Morley rank* by transfinite recursion:

(1) $MR(X) \geq 0$ if X is not empty;
(2) $MR(X) \geq \lambda$ if $MR(X) \geq \alpha$ for all $\alpha < \lambda$, for a limit ordinal λ;
(3) $MR(X) \geq \alpha + 1$ if there is an infinite family X_i of disjoint definable subsets of X such that $MR(X_i) \geq \alpha$ for all i.

Then we set $MR(X) = \sup\{\alpha : MR(X) \geq \alpha\}$, with the convention $MR(\emptyset) = -1$ and $MR(X) = \infty$ if $MR(X) \geq \alpha$ for all ordinals α.

If $MR(X) = \alpha$, we let the *Morley degree* $Md(X)$ be the maximal length d of a decomposition $X = X_1 \sqcup \cdots \sqcup X_d$ into pieces of rank α.

The additivity properties of Morley rank and degree justify the analogy between rank and dimension, and degree and multiplicity.

Proposition 9.

(1) *If $X \subseteq Y$, then $MR(X) \leq MR(Y)$.*
(2) $MR(X \cup Y) = \max\{MR(X), MR(Y)\}$.
(3) *If X and Y are disjoint with $MR(X) \leq MR(Y)$, then*

$$Md(X \cup Y) = \begin{cases} Md(X) + Md(Y), & \text{when } MR(X) = MR(Y), \\ Md(Y), & \text{otherwise.} \end{cases}$$

For a type p which contains ranked formulae, pick a formula $\varphi \in p$ of minimal rank α, and of minimal degree among all the formulae of rank α in p. We then let $MR(p) = \alpha$, and $Md(p) = d$. Such a φ *determines* p as the only type over A which contains it and has at least rank α.

Proposition 10. *Given a formula $\varphi \in \mathcal{L}(A)$, we have:*

(1) $MR(\varphi) = \max\{MR(p) : p \in S(A), \varphi \in p\}$;
(2) $Md(\varphi) = \sum\{Md(p) : p \in S(A), \varphi \in p, MR(p) = MR(\varphi)\}$.

In ACF, Morley rank coincides with the usual notion of dimension.

Proposition 11. *For a definable set X in ACF,*

(1) $MR(X)$ *equals the Krull dimension of the Zariski closure of X.*
(2) *For an algebraic set X, $Md(X)$ is the number of irreducible components of top dimension of X.*

We say that a theory T is *totally transcendental* if every definable set has (ordinal-valued) Morley rank. On the other hand, T is *ω-stable* if for every countable A, $S_1(A)$ is countable.

Example 15. By the proposition, ACF is totally transcendental. But DLO is not ω-stable, since there is a continuum of Dedekind cuts over $\mathbb{Q}$.

Proposition 12. *A theory T is totally transcendental if and only if T is ω-stable.*

Proof. If T is totally transcendental, every type over A is determined by an $\mathcal{L}(A)$-formula, and so
$$|S_1(A)| \leq |\mathcal{L}(A) - \text{formulae}| = |A| + \aleph_0.$$
Conversely, assume T is not totally transcendental, and thus $MR(x = x) = \infty$. Starting from $X_\emptyset = \mathfrak{C}$, we build a binary tree of definable subsets $(X_s : s \in {}^{<\omega}2)$ of infinite MR such that $X_s \neq \emptyset$, $X_{si} \subset X_s$ for $i = 0, 1$, and $X_{s0} \cap X_{s1} = \emptyset$. Choose a countable set A containing parameters for all X_s, and it is clear that every path $\sigma \in {}^\omega 2$ determines a type $p_\sigma \in S_1(A)$ which contains all X_s, for $s \subset \sigma$. The p_σ are all different so $S_1(A)$ has continuum many elements. $\qquad\square$

Recall (Corollary 5), a (type-) definable set X is *strongly minimal* if it is infinite and every definable subset of X is either finite or cofinite.

Lemma 2. *A definable set X is strongly minimal if and only if $MR(X) = 1$ and $Md(X) = 1$.*

Definition 4. A *pregeometry* (or a *matroid*) (A, cl) is a set A together with a closure operator $cl : \mathbb{P}(A) \to \mathbb{P}(A)$ such that:

(1) for any $B \subseteq A$, $B \subseteq cl(B) = cl(cl(B))$;
(2) if $B \subseteq C \subseteq A$ then $cl(B) \subseteq cl(C)$;
(3) if $B \subseteq A$ and $b \in cl(B)$ there is a finite $B_0 \subseteq B$ such that $b \in cl(B_0)$;
(4) (Steinitz exchange) if $B \subseteq A$, $b \in cl(B \cup \{c\}) \setminus cl(B)$ then $c \in cl(B \cup \{b\})$.

Example 16. These are archetypal pregeometries:

(1) a set with a trivial closure;
(2) a vector space with the linear span as closure;
(3) an affine space with the affine span;
(4) a field with the (relative) algebraic closure.

If (A, cl) is a pregeometry, we say that a set $B \subseteq A$ is *independent* if $b \notin cl(B \setminus b)$ for any $b \in B$. A *basis* for a set C is an independent set $B \subseteq C$ such that $C \subseteq cl(B)$. The following is proved exactly the same way as for vector spaces.

Proposition 13. *Suppose (A, cl) is a pregeometry and $B \subseteq A$. Any maximal independent subset of B is a basis for B. Moreover, any two bases for B have the same cardinality.*

One of the most important instances of pregeometries in model theory arises from algebraic closure in strongly minimal sets.

Definition 5. Let X be a definable set with parameters $\bar{d}$. For $Y \subseteq X$, let

$$\mathrm{acl}_X(Y) = \mathrm{acl}(Y \cup \{\bar{d}\}) \cap X.$$

Proposition 14. *If X is strongly minimal, then (X, acl_X) is a pregeometry.*

Morley's Theorem is often considered the spark that started a major development of model theory and stability.

Theorem 22 (Morley). *If T is categorical in some uncountable cardinal, it is categorical in all uncountable cardinals.*

Sketch of proof. Suppose T is categorical in an uncountable cardinality. The main conceptual steps in the proof are:

(1) T is ω-stable.
(2) T has a prime model (a model which has an elementary embedding into any other model).

I. Tomašić

(3) T has a strongly minimal formula $\varphi(x)$ over the prime model.
(4) If M and N are models of T of the same uncountable cardinality, there
     is a partial elementary bijection between $\varphi(M)$ and $\varphi(N)$ (think of
     bases).
(5) We extend this to an isomorphism $M \cong N$.   □

The next theorem deals with the number of models in countable cardinality.

Theorem 23 (Baldwin–Lachlan). *If T is uncountably categorical but
not $\aleph_0$-categorical, then T has exactly $\aleph_0$ non-isomorphic models of size $\aleph_0$.*

We can illustrate the ever present duality between the *quantitative* and
qualitative analysis in model theory using the stated results. The theorems
of Morley and Baldwin–Lachlan give information about the number of isomorphism types of models of given size, which is a form of classification, as
expounded in Section 6, but such results do not allow a qualitative classification. In that other direction, there is a stream of results which allow us,
under certain model-theoretic assumptions, to identify a classical (algebraic
or combinatorial) structure present, as discussed further in Section 7. For
now, we can state a prototypical result.

Theorem 24 (Macintyre). *An infinite ω-stable field is algebraically
closed.*

5.3. Stability

We say that a theory T is:

(1) *λ-stable,* if $|A| \le \lambda$ implies $|S_1(A)| \le \lambda$;
(2) *stable,* if it is λ-stable for some λ;
(3) *superstable,* if it is λ-stable for all sufficiently large λ.

A formula $\varphi(\bar{x}, \bar{y})$ has the *order property* if there are tuples $\bar{a}_i, \bar{b}_i$, $i < \omega$
such that

$$\models \varphi(\bar{a}_i, \bar{b}_j) \text{ if and only if } i \le j.$$

Proposition 15. *A theory T is stable if and only if there is no formula
with the order property.*

Example 17. The following are unstable in view of the above proposition:

(1) DLO;
(2) RCF, and even $\text{Th}(\mathbb{R}, +, \cdot, 0, 1)$.

Example 18 (Stable theories). We give a (non-exhaustive) list of stable theories, the study of which lay at the forefront of model theory research for years:

(1) Divisible torsion-free Abelian groups are ω-stable;
(2) Modules over a fixed ring are stable;
(3) ACF is ω-stable (since totally transcendental);
(4) Differentially closed fields of characteristic 0 are ω-stable, and those of characteristic p are stable not superstable;
(5) Separably closed fields (of fixed finite degree of imperfection) are stable not superstable;
(6) Free groups are stable, non-superstable if non-commutative.

6. Classification theory

Let T be a complete theory. We write $I(T, \lambda)$ for the number of isomorphism classes of models of T of cardinality λ. The assignment $\lambda \mapsto I(T, \lambda)$ is called the *categoricity spectrum* of the theory T.

Morley's Theorem can be rephrased in this language.

Theorem 25. *For a countable theory T, if $I(T, \lambda) = 1$ for some uncountable λ, then $I(T, \mu) = 1$ for all uncountable μ.*

Morley conjectured that for a countable T, $I(T, \cdot)$ is monotonous on uncountable cardinals, i.e., $\aleph_0 < \lambda \leq \mu$ implies $I(T, \lambda) \leq I(T, \mu)$. Shelah devoted 15 years to completely resolving all aspects of this problem, including determining the possibilities for the categoricity spectrum and classifying theories according to whether they have a *structure* or a *non-structure* theory.

Theorem 26 (Shelah's Main Gap). *For every countable T, either $I(T, \lambda) = 2^\lambda$ for every uncountable λ, or $I(T, \aleph_\alpha) < \beth_{\omega_1}(|\alpha|)$ and every model of T can be characterised upto isomorphism by an invariant of countable depth.*

We used the notation for *beth-numbers*, defined by transfinite recursion as

$$\beth_\beta(\mu) = \mu + \sum_{\gamma < \beta} 2^{\beth_\gamma(\mu)}.$$

Shelah then posed the *Classification Problem,* to classify the theories T in a useful way, so that for suitable questions on the class of models of T the partition to cases according to the classification will be helpful. He gave a complete solution in terms of dichotomy theorems associated with (not) having one of the five model theoretic properties: being *stable, superstable,* having properties *dop, deep, otop.* The only result we are able to mention states that only a superstable T has a hope of having a "structure theory".

Theorem 27. *If T is not superstable, for uncountable λ, $I(T, \lambda) = 2^\lambda$.*

Morley's conjecture readily follows from Shelah's theory. Shelah proceeds to study the classification for uncountable T and for "abstract elementary classes". This is a complete *quantitative* classification of first-order theories, but it does not give insight into a *qualitative* description of structures.

7. Geometric Model Theory

7.1. *Zilber's trichotomy*

Zilber famously conjectured that a model of an uncountably categorical theory must (essentially) be either:

- Trivial (such as a set with no structure);
- A vector space;
- An algebraically closed field.

In order to identify which, we consider the types of pregeometries arising on strongly minimal sets as in Proposition 14, bearing in mind Example 16.
    A pregeometry (X, cl) is:

(1) *Trivial,* if $cl(A) = \cup_{a \in A} cl(\{a\})$ for all $A \subseteq X$;
(2) *Modular,* if whenever $A, B \subseteq X$ are finite dimensional,
$$\dim(A) + \dim(B) = \dim(A \cup B) + \dim(A \cap B);$$

(3) *Locally modular* if the above holds whenever $A \cap B \neq \emptyset$.

In these terms, Zilber's principle would entail that a non-locally modular strongly minimal set "interprets" an algebraically closed field.

In a similar direction, a conjecture of Cherlin and Zilber states that an infinite simple group of finite Morley rank must be an algebraic group over an algebraically closed field.

We can now list some (qualitative) classification results.

Theorem 28 (Cherlin–Harrington–Lachlan, Zilber).

(1) *An ω-categorical strongly minimal set is locally modular.*
(2) *An ω-categorical strictly minimal set (where $cl(a) = \{a\}$ for all $a \in X$) is either a pure set, or a projective or affine geometry over a finite field.*

Using a sophisticated variant of Fraissé amalgamation, Hrushovski refuted Zilber's conjecture in full generality by producing a non-locally modular strongly minimal type which does not even interpret a group. Hrushovski and Zilber then prove that under the assumption on the existence of a certain topology on a structure (reminiscent of the Zariski topology in algebraic geometry), Zilber's trichotomy holds. They dub such structures *Zariski geometries*.

The Cherlin–Zilber conjecture is still open, even though many cases have been dealt with and it is now widely believed to be true. The methods used are a mixture of techniques for studying *algebraic groups* and the ones employed in the *classification of finite simple groups* through the analogy between "finite" and "finite rank".

8. Model Theory and Diophantine Geometry

Some of the most spectacular applications of model theory in diophantine geometry came through Hrushovski's proofs of Mordell–Lang and Manin–Mumford conjectures using stability theory, and Pila–Zannier strategy to prove André–Oort conjecture using o-minimality. Although the details are way beyond the scope of these notes, we hope to give a summary of these methods at a conceptual level, and show how they fit with the model theory we covered. This section is meant to recall the necessary background from diophantine geometry and number theory, inspire and give directions for further reading.

8.1. *Abelian varieties*

An *Abelian variety* is a connected complete algebraic group.

Example 19 (Elliptic curves). An *elliptic curve* is a smooth projective genus one curve endowed with a point O. For example, a complex cubic E

given by the equation $zy^2 = x^3 + axz^2 + bz^3$ in $\mathbb{P}^2$, where $4a^3 + 27b^2 \neq 0$. Since E is a cubic, every line intersects it in 3 points (with multiplicities). For $P, Q \in E(\mathbb{C})$, let $P * Q$ denote the third point of intersection of the line PQ with E (for $P = Q$ take the tangent at P). The rule $P \oplus Q = O * (P * Q)$ defines an Abelian group structure on $E(\mathbb{C})$ with identity O, making E into a projective algebraic group, and thus an Abelian variety.

The j-invariant of E is defined as $j = 1728\frac{4a^3}{4a^3+27b^2}$. Two elliptic curves are isomorphic if and only if their j-invariants are equal, allowing us to consider $\mathbb{A}^1$ as the moduli space of elliptic curves. Indeed, a point in $\mathbb{C} = \mathbb{A}^1(\mathbb{C})$, thought of as a j-invariant, encodes an isomorphism class of elliptic curves.

Theorem 29. *Let A be an Abelian variety over an algebraically closed field k. Then A is commutative (which explains the name), divisible and torsion points are Zariski dense. In characteristic 0, $A_{torsion} = (\mathbb{Q}/\mathbb{Z})^{2g}$, and in arbitrary characteristic, if $l \neq char(k)$, $\ker[l^n] = (\mathbb{Z}/l^n\mathbb{Z})^{2g}$.*

Each smooth and projective curve X of genus $g \geq 1$ can be embedded in its *Jacobian variety* $J(X)$ which is an Abelian variety of dimension g classifying degree zero divisors up to rational equivalence. Namely, if P_0 is a point on X, the embedding $X \to J(X)$ can be thought of as $P \mapsto [P] - [P_0]$.

For an elliptic curve E, $J(E)$ is isomorphic to E.

8.2.   *Beginnings of diophantine geometry*

Theorem 30. *Let X be a smooth and projective curve over $\mathbb{Q}$. Then:*

(1) If X is of genus 0, then either $X(\mathbb{Q}) = \emptyset$ (e.g., $x^2 + y^2 + 1 = 0$), or all but finitely many solutions are parameterised by rational fractions (e.g., all solutions of $x^2 + y^2 - 1 = 0$ except $(0, 1)$ are parametrised by $(2t/(t^2 + 1), (t^2 - 1)/(t^2 + 1))$).

(2) If X is of genus 1, then either $X(\mathbb{Q}) = \emptyset$ or X is an elliptic curve, so Mordell–Weil Theorem states that $X(\mathbb{Q})$ is a finitely generated group. More generally, for an Abelian variety A and K a number field, $A(K)$ is finitely generated.

(3) If X is of genus ≥ 2, then Faltings' Theorem *(originally* Mordell conjecture*) states that $X(\mathbb{Q})$ is finite.*

We wish to give restatements of the above results suitable for generalisations. Let X be a curve in an Abelian variety A.

For *Mordell's conjecture for curves,* let Γ be a finitely generated subgroup of A (such as $A(\mathbb{Q})$).

Then $X \cap \Gamma$ is finite, except when X is a translate of an elliptic curve.

This can be proved by combining Mordell–Weil and Faltings' theorems, and by observing X in its Jacobian $A = J(X)$.

In *Manin–Mumford conjecture for curves,* we wish to have the same conclusion for $\Gamma = A_{\text{torsion}}$.

Lang's conjecture for curves refines both Mordell's and Manin–Mumford's, by requiring the same conclusion for Γ a subgroup of finite rank (divisible hull of a finitely generated subgroup) of A.

8.3. *Diophantine equations over function fields*

If k is an algebraically closed field, a *function field* K over k (of transcendence degree 1) is the field of rational functions of a variety (of dimension 1) over k.

An old adage in number theory states that there should be a deep analogy between number fields and function fields, so we should have function field versions of all the above results. However, given a curve X over K, the natural question to ask is not whether $X(K)$ is finite, but possibly whether $X(K) \setminus X(k)$ is finite, as the following example shows.

Example 20. Let X be the Fermat curve defined by $X^n + Y^n = Z^n$ for some $n \geq 3$. Then $X(\mathbb{C}(T)) \setminus X(\mathbb{C}) = \emptyset$.

Theorem 31 (Mordell's conjecture over function fields). *Let X be a curve of genus ≥ 2 defined over K which is a function field over k. Then $X(K)$ is finite unless X is* isotrivial, *i.e., there is a curve X_0 defined over k and isomorphic to X over some finite extension K' of K.*

There is also a *function field/relative Mordell–Weil* theorem. It is easiest to state when A is an Abelian variety over K with *K/k-trace 0,* i.e., with no non-zero homomorphic images defined over k, in which case it simply states that $A(K)$ is finitely generated.

We will consider further relative versions of diophantine statements in the next subsection.

8.4. *Diophantine conjecture template*

Suppose we are given an abstract context consisting of:

- A variety A of a certain type, usually of significance in diophantine geometry;
- A subset Γ of *special points* of A;
- A collection of *special subvarieties* of A.

Template Theorem. If special points are Zariski dense in an irreducible subvariety X of A, then X must be special.

Note that the restatements for curves from the previous subsection fit this template perfectly, in view of the fact that a set is Zariski dense on a curve if and only if it is infinite. On the other hand, we can exploit this template to give higher-dimensional analogues of these conjectures as follows.

(1) *Manin–Mumford (MM).* In the template, A is an Abelian variety over $\mathbb{C}$, special points are torsion points ($\Gamma = A_{\text{torsion}}$), special subvarieties are translates of Abelian subvarieties by torsion points.

(2) *Mordell.* Again, A is Abelian over $\mathbb{C}$, special points are elements of a finitely generated subgroup Γ of A, special subvarieties are translates of Abelian subvarieties by elements of Γ.

(3) *Lang (ML).* As above, with Γ a finite-rank subgroup of A.

(4) *Mordell–Lang function field case all characteristics (relative ML).* Let K/k be an extension of algebraically closed fields, A an Abelian variety over K, and Γ a finite rank subgroup of A. A subvariety X of A is special, if there exists an Abelian subvariety S of A, an Abelian variety S_0 over k, a subvariety X_0 of S_0 over k, and a rational homomorphism $h : S \to S_0$ so that X is a translate of $h^{-1}(X_0)$.

When the K/k-trace of A is 0, the definition of special subvariety simplifies to just "a translate of an Abelian subvariety of A".

(5) *André–Oort conjecture (AO).* In this case, A is a *(mixed) Shimura variety*, and "special" points and subvarieties have a technical meaning in that context. It is not possible to give a definition of Shimura varieties here, but readers can think of examples such as moduli spaces $\mathcal{A}_g$ of (principally polarised) Abelian varieties of dimension g (we already discussed $\mathcal{A}_1 \simeq \mathbb{A}^1$ in Example 19) and their products.

8.5. Model-theoretic proofs

Definition 6. Let K be an algebraically closed field and A a commutative algebraic group over K (which we identify with its set of K-rational points) and let Γ be a subgroup of A.

We say that the triple (K, A, Γ) is of *Lang-type* if for every n and every subvariety X (over K) of A^n, $X \cap \Gamma^n$ is a finite union of cosets.

Remark 2. If Γ is a finite rank subgroup of a semi-Abelian variety A over $\mathbb{C}$, Lang's conjecture is equivalent to the statement that $(\mathbb{C}, A, \Gamma)$ is of Lang-type.

The non-trivial implication is obtained by considering the Zariski closures of the relevant cosets. A similar remark can be made in the relative case.

Model theory considerations enter through the observation that Lang-type behaviour is characteristic of *one-based stable groups*.

Theorem 32 (Hrushovski–Pillay). *Let T be a stable theory, M a big model of T and G an $\emptyset$-definable group in M. Then G is one-based in T (a property generalising modularity mentioned before in the strongly minimal setting) if and only if every definable (with parameters) subset of G^n is a finite Boolean combination of cosets (of definable subgroups of G^n).*

Corollary 9. *The structure (K, A, Γ) is of Lang-type if and only if $\mathrm{Th}(K, +, \cdot, \Gamma, a)_{a \in K}$ is stable and the formula "$x \in \Gamma$" is one-based.*

Sketch of proof of MM, and relative ML in all characteristics. In all cases, we would like to show the respective conclusion by proving that $\mathrm{Th}(K, +, \cdot, \Gamma, a)_{a \in K}$ is stable and Γ is one-based, but the problem is that Γ is not definable in any natural sense of the word.

Thus we must replace Γ by a *definable* object which is small enough (finite rank) so that the conclusion will still hold. The language of rings is too coarse, so we expand the language as follows.

- In ML characteristic 0, we replace Γ by a definable group of finite Morley rank in the language of *differential fields* (we add a symbol for the derivation such that k is contained in the constants).
- In ML characteristic p, we replace K by a *separably closed* but not algebraically closed field K and use the fact that the definable sets are much finer than just constructible sets, and Γ is replaced by a type-definable object of finite U-rank.

- In the proof of MM, the torsion is expressed using the language of *difference fields* (fields with a symbol for a distinguished endomorphism), which are not stable, but are *simple*, i.e., still have a good theory of forking and some local ranks.

   The problem reduces to the known model-theoretic analysis of differential, separably closed and difference fields where the non-locally modular phenomena are "tied" to the field of constants, K^{p^∞} and the fixed field, respectively.   □

Idea of proof of most cases of AO. We find an analytic covering $\pi : D \to A(\mathbb{C})$, where D is a somewhat bounded domain, so that, identifying $\mathbb{C}$ with $\mathbb{R} \times \mathbb{R}$, π becomes definable in the o-minimal structure $\mathbb{R}_{an,exp}$, which is an expansion of $\mathbb{R}$ by the exponential function and restrictions of all analytic functions to a bounded interval. In this situation, counting special points on $X \subseteq A$ reduces to counting rational points on $\pi^{-1}(X)$ in some number field. The conclusion follows by playing off the upper bounds for the number of rational points on sets definable in o-minimal structures by Pila–Wilkie against the lower bounds from the estimation of sizes of Galois orbits from number theory, sometimes requiring the generalised Riemann hypothesis.

   In our example of $A = \mathcal{A}_1$, we exploit the analytic description of elliptic curves as complex tori. Each element τ in the upper half plane $\mathfrak{h} = \{z \in \mathbb{C} : \Im(z) > 0\}$ defines an elliptic curve $E_\tau = \mathbb{C}/(\mathbb{Z} + \mathbb{Z}\tau)$. The group $\mathrm{PSL}_2(\mathbb{R})$ acts transitively on $\mathfrak{h}$ by fractional linear transformations $\left(\begin{smallmatrix} a & b \\ c & d \end{smallmatrix}\right) \cdot z = \frac{az+b}{cz+d}$, and E_σ and E_τ are isomorphic if and only if σ and τ are in the same orbit of the modular group $\mathrm{PSL}_2(\mathbb{Z})$, so our moduli space $\mathcal{A}_1$ can be identified with $\mathrm{PSL}_2(\mathbb{Z})\backslash\mathfrak{h}$. The analytic j-function $j : \mathfrak{h} \to \mathbb{C} = A(\mathbb{C})$ is a surjective holomorphic function which is exactly $\mathrm{PSL}_2(\mathbb{Z})$-invariant, so $E_\sigma \simeq E_\tau$ if and only if $j(\sigma) = j(\tau)$. The restriction of j to the fundamental domain $D = \{\tau \in \mathfrak{h} : |\tau| \geq 1, -1/2 \leq \Re(\tau) \leq 1/2\}$ provides the required analytic covering π. Special points in A are values $j(\tau)$ for which E_τ has *complex multiplication*, which happens exactly for quadratic imaginary τ.   □

9.   Further Reading

The basic model theory part of these notes heavily relies on Marker's masterful survey [9]. For more on model theory, we recommend his book [10], as well as [2] and [5] for the lovers of classics.

For stability and geometric stability, we recommend [11]. For ambitious readers with interest in classification, there is Shelah's masterpiece [13].

The classic reference for algebraic geometry is [4]. For p-adic integration, we suggest [3]. Our survey of diophantine geometry is based on [1] for material on MM and ML, but we also recommend the original papers [7], [6]. The work on AO is still in progress (authors including André, Daw, Edixhoven, Habegger, Klingler, Moonen, Orr, Peterzil, Pila, Starchenko, Tsimerman, Ullmo, Yafaev), and we followed the survey [12].

Acknowledgements

The author would like to thank Angus Macintyre, Anand Pillay and Andrei Yafaev for helpful discussions on various aspects of these notes.

10. Exercises

1. Find a first-order axiomatisation for the class of torsion-free divisible Abelian groups and prove that this theory is complete. Prove that the property "G is torsion" cannot be axiomatised in a first-order way in the language of groups.

2. Working in the theory ACF, eliminate the quantifiers from:
$$\exists w \, \exists x \, \exists y \, \exists z (aw + by = 1 \, \wedge \, ax + bz = 0 \, \wedge \, cw + dy = 0 \, \wedge \, cx + dz = 1).$$

3. Is the theory ACF_p $\aleph_0$-categorical? Is DLO uncountably categorical?

4. Show the existence of a *countable* non-standard (non-isomorphic to $\mathbb{N}$) model of PA. Show the existence of a non-archimedean model of RCF.

5. Let θ be an $\forall\exists$-sentence, i.e.,
$$\theta \equiv \forall x_1 \ldots \forall x_n \exists y_1 \ldots \exists y_m \, \varphi(x_1, \ldots, x_n, y_1, \ldots, y_m),$$
where φ is quantifier-free. Let M_i, $i \in I$ be a chain of structures indexed by a linear order $(I, <)$ such that $M_i \models \theta$ for all $i \in I$. Prove that $\bigcup_{i \in I} M_i \models \theta$.

6. Let $\sigma : \mathbb{C}^n \to \mathbb{C}^n$ be an algebraic automorphism of $\mathbb{C}^n$ viewed as an algebraic variety (the affine n-space $\mathbb{A}^n$). In other words, the components of σ are polynomial maps. Prove that, if $\sigma^2 = 1$, then σ has a fixed point.

7. Use model-completeness of DLO to prove that *order-completeness* (the property that every non-empty subset with an upper bound has a supremum) is not expressible in first-order logic.

8. Let F be a real closed field. An ideal $I \subseteq F[x_1, \ldots, x_n]$ is *real*, if $f_1^2 + \cdots + f_m^2 \in I$ implies $f_1, \ldots, f_m \in I$. Formulate and prove analogues of Theorem 16 for real prime ideals, and Proposition 7 for RCF.

9. Verify that countable models of DLO are $\aleph_0$-saturated. What is the minimum size for an $\aleph_1$-saturated model?

10. Verify the statement of Example 14.

11. It is well-known in algebra that $\mathbb{C}$ cannot be made into an ordered field. Prove that it cannot even be made into a total order by a first-order formula in the language of rings.

12. For a topological space X and an ordinal α, find the definition of the α-*th Cantor–Bendixson derivative* X^α on Wikipedia. We apply these considerations to the Stone space $X = S_n(M)$ of an $\aleph_0$-saturated model M of a complete theory T. Given a type $p \in X$, we say that its *Cantor–Bendixson rank* is α, written $CB(p) = \alpha$, if $p \in X^\alpha \setminus X^{\alpha+1}$. Prove that:

(a) $MR(p) \geq CB(p)$;
(b) if T is totally transcendental, $MR(p) = CB(p)$.

10.1. *Solutions and hints to selected exercises*

2. Performing the known QE algorithm would be laborious. Instead, note that the formula expresses the fact that the matrix $\left(\begin{smallmatrix} a & b \\ c & d \end{smallmatrix}\right)$ is invertible, which is equivalent to the condition $\neg(ad - bc = 0)$.

3. No to both. Algebraic closures of $\mathbb{Q}$ and $\mathbb{Q}(x)$ are countable, but $\mathbb{Q}(x)$ does not embed into $\bar{\mathbb{Q}}$ since x is not algebraic over $\mathbb{Q}$. Regarding DLO, either argue that it is not uncountably categorical because it is not stable (by the first step in the proof of Morley's Theorem), or directly construct non-isomorphic models in an uncountable cardinality.

4. Add a constant c to the language of PA and consider the set of axioms PA $\cup \{\underbrace{1 + \cdots + 1}_{n \text{ times}} < c : n \in \mathbb{N}\}$. Any finite subset of those axioms is consistent (it is satisfied by a model of PA in which c is interpreted as some number larger than any n mentioned in the finite subset), so the compactness/completeness and the Löwenheim–Skolem theorems yield a countable model in which c is greater than any standard natural number. The RCF question can be done in a similar way.

6. Note that the statement holds over finite fields of odd characteristic (not all σ-orbits can be of size two) and conclude by the Lefschetz principle in the spirit of Ax's Theorem 9.

7. Since $(\mathbb{Q}, <), (\mathbb{R}, <) \models$ DLO, by model-completeness we get $\mathbb{Q} \preceq \mathbb{R}$. On the other hand, $\mathbb{R}$ is order-complete, and $\mathbb{Q}$ is not.

8. Hint: think of Dedekind cuts, there are continuum many.

11. Stability of ACF shows that there is no formula with the order property.

References

[1] E. Bouscaren, editor, Model theory and algebraic geometry, *Lecture Notes in Mathematics* **1696**, Springer-Verlag, Berlin, 1998.

[2] C. C. Chang and H. J. Keisler, Model theory, *Studies in Logic and the Foundations of Mathematics* **73**, North-Holland Publishing Co., Amsterdam, third edition, 1990.

[3] J. Denef, Arithmetic and geometric applications of quantifier elimination for valued fields, in *Model theory, algebra, and geometry, Math. Sci. Res. Inst. Publ.* **39**, pp. 173–198, Cambridge University Press, Cambridge, 2000.

[4] R. Hartshorne, *Algebraic Geometry*, Graduate Texts in Mathematics, **52**, Springer-Verlag, New York-Heidelberg, 1977.

[5] W. Hodges, Model theory, *Encyclopedia of Mathematics and its Applications* **42**. Cambridge University Press, Cambridge, 1993.

[6] E. Hrushovski, The Mordell-Lang conjecture for function fields, *J. Amer. Math. Soc.*, **9(3)**: 667–690, 1996.

[7] E. Hrushovski, The Manin-Mumford conjecture and the model theory of difference fields, *Ann. Pure Appl. Logic*, **112(1)**: 43–115, 2001.

[8] S. Lang, Algebra, *Graduate Texts in Mathematics*, **211**, Springer-Verlag, New York, third edition, 2002.

[9] D. Marker, Introduction to model theory, pp. 15–35 in *Model theory, algebra, and geometry, Math. Sci. Res. Inst. Publ.* **39**, Cambridge University Press, Cambridge, 2000.

[10] D. Marker, *Model theory, Graduate Texts in Mathematics*, **217**, Springer-Verlag, New York, 2002.

[11] A. Pillay, Geometric stability theory, *Oxford Logic Guides* **32**, The Clarendon Press, Oxford University Press, New York, 1996.

[12] T. Scanlon, A proof of the André-Oort conjecture via mathematical logic [after Pila, Wilkie and Zannier], *Astérisque*, (348):Exp. No. 1037, ix, 299–315, 2012.

[13] S. Shelah, Classification theory and the number of nonisomorphic models, *Studies in Logic and the Foundations of Mathematics* **92**. North-Holland Publishing Co., Amsterdam, second edition, 1990.

Printed in the United States
By Bookmasters